TUNING FOR SPEED
AND
TUNING FOR ECONOMY

BOOKS BY PHILIP H. SMITH
PUBLISHED BY ROBERT BENTLEY, INC.

Maintenance, Tuning and Repair.

    Tuning & Maintenance of MGs.
    Tuning for Speed & Tuning for Economy.

Design and Engineering.

    The Design & Tuning of Competition Engines.
    Valve Mechanisms for High-Speed Engines.
    The Scientific Design of Exhaust and Intake Systems.

Tuning for Speed
and
Tuning for Economy

by

PHILIP H. SMITH

C.Eng., F.I.Mech.E., M.S.A.E.

Fourth Edition edited and revised by
T. C. MILLINGTON, B.A.

Robert Bentley, Inc.
872 Massachusetts Avenue
Cambridge, Massachusetts 02139

First Impression July, 1955
Second Impression September, 1955
Revised Edition January, 1962
Second Impression January, 1964
Third Impression April, 1965
Revised Third Edition, 1967
Revised Fourth Edition, 1970
Second Impression August, 1972
Third Impression August, 1974
Fourth Impression September, 1975

© **Philip H. Smith**
Library of Congress Catalog Card No. 72-86570
ISBN 0-8376-0005-7

Manufactured in the United States of America

CONTENTS

	Page
LIST OF DRAWINGS AND DIAGRAMS	vii
INTRODUCTION	ix

Chapter

1. **VALVES AND VALVE GEAR** 1
   Clearance measurement—Side-valve engines—o h c engines—Lubrication of o h v gear—Oil-tightness

2. **SPARK PLUGS** 12
   Spark plug welfare—Plug examination—Plug cleaning—Non-detachable plugs—Replacing in the engine—Sparking devices

3. **IGNITION EQUIPMENT** 19
   The contact-breaker mechanism—Modern wiring—The make-and-break—The automatic advance mechanism—Gap setting—Final assembly—Timing for efficiency—The distributor position—Road testing—Optimum setting

4. **THE INDUCTION SYSTEM AND CARBURETTER** 37
   The induction pipe—Freedom from air leaks—The carburetter—The SU principle—Examination of the moving parts—Rectification of faults—Assembling the dashpot—Complications with the jet—Reassembling the jet—The float-chamber assembly—Fuel level adjustment—Spring return dashpots—SU carburetter adjustment—Starting mixture—Diaphragm-jet type—Control linkage—Special adjustments—Auxiliary starting carburetter—Starter control—The Stromberg CD carburetter—Starting and idling—Normal running—Adjustments—Float level—Constructional points—fixed-choke carburetters—Solex carburetter—Starting mixture—Adjustment of the Solex carburetter—Selection of part sizes—Examples of jet selection—General maintenance—The Zenith carburetter—Working principle—Zenith adjustment—Slow running—Faults and their causes—Multiple carburetters—Aural check—Reliability

5. **FUEL SUPPLY APPARATUS** 85
   The SU electric pump—Delivery to order—Contact-breaker—Dismantling the pump—The valves—Checking the valves—The diaphragm—Stretching the diaphragm—Testing the pump—The contact-breaker assembly—Adjusting the points—The PD electric pump—Filter cleaning—Faulty action—The AC mechanical pump—Overhauling—Diaphragm refitting—Fuel connections

CONTENTS

6. **THE LUBRICATION SYSTEM** 108
   Engine lubrication—Oil pressure—Other lessons from the gauge—Filters—Low pressure—Other causes of low pressure—Bearing wear—Auxiliary bearings—Choice of lubricants—Thin oil advantages—Draining the sump—Under-sump drain plugs—Flushing the sump—Renewable-element filters—Care in assembly—Throw-away filters—Fitting a by-pass filter—The connecting pipes

7. **THE COOLING SYSTEM** 131
   The cooling water—Flushing—Leakage—Hose clips—The thermostat

8. **ELECTRICAL MATTERS** 139
   The electrical system—General principles—Live wiring—Circuit fuses—Current capacity—No-fuse wiring—Wiring and accessories—Soldering aids—Other connections—Cable runs—Wiring security—Insulation of cables—Earthing

9. **SEASONAL WELFARE** 151
   Climatic conditions—Other handicaps—About the battery—Methodical charging—Generator and wiring—The starter motor—Bearings—Sticking pinion—Ignition ills—Extra insulation circuit—High-tension circuit—Water ingress—The carburetter—Fuel pipes—'Taxi' operation—Radiator connections—The thermostat—Filtration and cooling—The ignition system—Vapour locks

10. **ACCENT ON ECONOMY** 169
    First, the driver—Engine efficiency—Exhaust loss—Limitations—Carburetter and distribution—Carburetter matters—Variety of devices—Automatic Air valves—Good idling—Opinion—Economy jets—Restricted breathing—Air filters

11. **A CHAPTER ON GADGETS** 182
    Boosted ignition—Low-tension matters—Resetting of plugs—Spark-gap terminals—Distributor details—Extra air—Mixture agitators—Intake silencers—Special intakes—Alternative filters—Carburetter changes—Exhaust pipe fittings

12. **EMISSION CONTROL SYSTEMS** 194
    Sealed breathing—Exhaust emissions—Clean Air system—Manifold Air Oxidation—Duplex manifold system—Other systems—Evaporative emissions

13. **POWER TO THE BEST ADVANTAGE** 201
    Legitimate power absorption—Saving power—Brake adjustment—Useful tests—Test-tune equipment—The equipment—Interpretation of readings—Gauge readings—Conclusion

    APPENDIX 211
    Explanatory definitions—Car performance

    INDEX 215

Illustrations

Adjusting rocker clearance	page 1
Measuring rocker clearance	2
Single-nut rocker adjustment	4
Self-locking tappet	5
Position of cam for maximum clearance	6
Measuring clearance on o h c	7
Valve stem oil seal	9
Typical spark plugs	13
Gapping the plugs	17
High tension cable terminal	20
HT cable secured from inside the cover	21
Old type contact-breaker	24
Components of distributor with centrifugal governor	25
Components of distributor with centrifugal- and vacuum-control	26
'Quikafit' contacts	27
Contact-breaker with two-screw fixing	29
Contact-breaker with single-screw fixing	30
Lubricating the cam and contact pivot	32
Timing marks	34
Principle of SU carburetter	42
Section through SU carburetter	43
Section through SU carburetter	44
Cleaning and oiling SU carburetter	45
Fitting needle to SU carburetter	48
Centralising the jet (SU)	50
Adjusting the mixture (SU)	51
Jet assembly of SU carburetter	52
Filter and union (SU)	54
Adjusting the fuel level (SU)	55
Adjusting the fuel level (SU)	56
Position of adjusting screw (SU)	58
Sideview of diaphragm-jet carburetter	59
End view of diaphragm-jet carburetter	60
Stromberg CD carburetter	65
Solex 32B1 carburetter	70
Zenith V-type carburetter	77
Section through SU electric fuel pump	page 87
Components of electric fuel pump	89
Details of contact-breaker assembly, SU fuel pumps	91
Lining up contact points of fuel pumps	92
Adjusting the diaphragm spindle	95
Order of assembly, pump terminal	97
Dressing the points, fuel pump	98
PD pump on Mini	100
Components of PD pump	101
Section through AC pump	103
Components of AC pump	105
A lubrication system	109
Section through full-flow filter	113
Sump drain plug	120
Riley oil filter	123
Section through throw-away filter	123
Hillman Imp oil filter	124
Throw-away filter on MG	127
Radiator and cylinder block drain taps	133
Hosepipe adaptor	134
Section through radiator cap	134
Sectional view of water-pump	135
Lubrication of water-pump	136
Fitting hose clips	137
Thermostat components	138
Electrical control box	142
Mini fuse-block	143
A variety of wiring terminals	145
Lubricating the generator bearing	155
Freeing a jammed starter	159
Hot spot with flap valve	173
Gauze air cleaner	188
Two air-filters	190
Exhaust tail-pipe fittings	192
Stromberg CDSE carburetter	196
Manifold-Air-Oxidation	197
Duplex manifold system	199
Early type brake adjusters	203
Brake adjustment through hole in drum	204
Brake adjuster on backplate	205
Disc brake assembly	205
Free pedal movement	206

vii

ACKNOWLEDGEMENT

My thanks are due to the following manufacturers for their kind co-operation in allowing the use of illustrations.
The British Motor Corporation Ltd
The Ford Motor Co Ltd
Joseph Lucas Ltd
Rootes Motors Ltd
Solex Limited
The S U Carburetter Co Ltd
Zenith Ltd

NOTE

No reference is made in this book to matters concerning intricate adjustments to the electrical generating equipment. It is considered that such equipment, which usually conforms to an established pattern, demands specialised attention. On any occasion when trouble is experienced therefore, recourse to the maker's service station will usually prove the best policy.

Introduction

This book can be regarded as 'two books in one'; it is a fact that the two desirables, speed and economy, are inseparably related in an engine that is operating efficiently, since it will be producing its intended power with as little fuel wastage as possible. Two popular misconceptions, however, are still occasionally found; first that by extravagant petrol usage it is possible to obtain a considerable power boost; second, that by doing things at the carburetter (plus perhaps the use of some additional 'petrol-saver' fitting) the mileage per gallon can be increased with little effect on performance.

These pages give no encouragement to either attitude. The answer to economical motoring is to choose the size of car on this basis; the recipe for speed is correct tune. For those who for any reason find it necessary on an existing vehicle to undergo the penance of trying to save fuel, Chapter 10 will be of benefit.

There are few engines to standard specification that are not up to their job so long as their tune is right. However, those who require better-than-standard performance, to the extent of being prepared to carry out considerable modifications, will find much of interest in the Author's companion book *Car Performance and the Choice of Conversion Equipment*.

The average driver is not so insensitive to defects in performance as some of the more enthusiastic types imagine: furthermore, he is usually quite capable of carrying out his own running adjustments, most of which require quite a lot of time to do properly. Which brings us to that overworked word 'tuning' which by tradition should not be, though it now is, applied to structural modifications to the power unit. In the old days, even racing enthusiasts usually had to make do with the engine as delivered, and their tuning was of the right persuasion—meticulous attention to detail. If we agree for once that the word implies simply that everything is operating in tune to maximum advantage, we can see immediately that the art need by no means be confined to the owner of a sporting type of vehicle, but is open to the family motorist as well, in equal measure.

CHAPTER 1

Valves and Valve Gear

*Clearance measurement—Side-valve engines—Lubrication of o h v gear—
o h c engines—Oil-tightness*

For good engine performance the valve mechanism must carry out its function properly, and correct adjustment of the working clearance is essential. The conventional type of o h v engine with push-rod and rocker valve operation has the adjustment point located on the rocker-end at the top of the push-rod, and requires the use of a screwdriver for the adjustment screw and a spanner for the lock-nut as shown on Fig. 1.1. The clearance between the valve-stem tip and the hardened end of the rocker is measured by a feeler gauge as shown on Fig. 1.2.

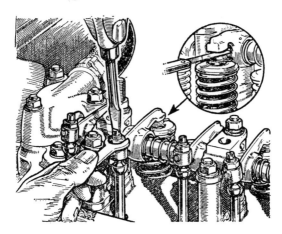

Fig. 1.1. Adjusting rocker clearance on Wolseley 15/60 engine. Inset shows feeler gauge.

There are several points to bear in mind when undertaking this task. First, in regard to the clearance specified, this may be stated in so many thousandths of an inch, 'hot' and 'cold'. It is, however, preferable to use the figure quoted for the hot engine, unless the makers definitely specify otherwise. The reason for this is that the various components may vary slightly in their rates of expansion, so that what was a regular clearance on all the rockers when set cold, might well vary as between different rockers at running temperature.

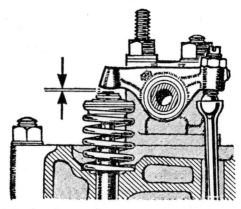

Fig. 1.2. Rocker clearance is measured at the gap shown arrowed.

And the running temperature must be, in fact, just that. A short run in the garage is no use; the engine must be really warm, as after a fast bout of road work, when water and oil will be at the right temperature.

Clearance measurement

When an engine has covered a reasonable mileage, the end of the valve rocker which bears on the valve stems takes on a minute indentation, and is thus no longer flat. Thus, the measurement of clearance by feeler gauge requires some care. The only infallible method is that of 'go' and 'not go'. For example, if the specified clearance is 0.015 in., a feeler of this value should be inserted and the adjustment made by the screw until the feeler is just trapped in the clearance between the valve-stem and the rocker tip. The lock-nut is then lightly tightened. Now withdraw the feeler and insert one to the value of 0.014 in. This should go in quite easily. For the 'not go' test we use a 0.016 in. feeler which should not be capable of being inserted at all. If the adjustment passes this test, the lock-nut may be fully tightened, after which the whole procedure must be repeated in case this tightening has altered the adjustment.

This procedure holds good for all engines, the rule being to try three feelers, one having the specified clearance, and the other two being respectively one 'thou' above and below that. It may be that the pack of gauges used does not contain single feeler blades in the sizes required, but combinations of blades will be found to give any desired value. When using multiples of blades in this manner, always use as few as possible to make up the required thickness, and keep them clean.

Valve clearance has, of course, to be measured with the valve fully closed, which means that the tappet must be on the back of its cam. Some makers provide a table indicating that when a certain valve is fully open (which is easily seen) a corresponding valve is fully closed. For example, this procedure for a typical four-cylinder engine firing 1.3.4.2, is as follows:—

Adjust No. 1 rocker with No. 8 valve fully open.
,, ,, 3 ,, ,, ,, 6 ,, ,, ,,
,, ,, 5 ,, ,, ,, 4 ,, ,, ,,
,, ,, 2 ,, ,, ,, 7 ,, ,, ,,
,, ,, 8 ,, ,, ,, 1 ,, ,, ,,
,, ,, 6 ,, ,, ,, 3 ,, ,, ,,
,, ,, 4 ,, ,, ,, 5 ,, ,, ,,
,, ,, 7 ,, ,, ,, 2 ,, ,, ,,

In the case of a six-cylinder in-line engine firing 1.5.3.6.2.4, the sequence is as under:

Adjust nos 1 and 3 rockers with nos 10 and 12 valves fully open
,, ,, 8 ,, 11 ,, ,, ,, 2 ,, 5 ,, ,, ,,
,, ,, 4 ,, 6 ,, ,, ,, 7 ,, 9 ,, ,, ,,
,, ,, 10 ,, 12 ,, ,, ,, 1 ,, 3 ,, ,, ,,
,, ,, 2 ,, 5 ,, ,, ,, 8 ,, 11 ,, ,, ,,
,, ,, 7 ,, 9 ,, ,, ,, 4 ,, 6 ,, ,, ,,

The engine may be rotated by the starting handle if fitted, or by other means such as pushing the car in gear or pulling on the fan blades. An alternative method of adjustment is to turn the engine until the valve being dealt with is fully open, after which it is rotated through a further complete revolution. This will of course result in the camshaft turning through a half-revolution, bringing the valve to its fully-closed position for adjustment.

It is preferable to remove the sparking plugs when adjusting valve clearances; this takes little extra time which is well repaid in the greater ease of turning the engine, and the precision with which the exact position of rotation can be arrived at. It should also be noted that some makers may object to the fan blades being pulled round for the purpose, as the balance of the fan might be affected by bending. This point should be checked with the makers before using the method.

The final requirement in this adjustment is to lock up the nut on the rocker really tightly, after obtaining the final setting. A stout and well-fitting spanner is essential; any makeshifts such as adjustable wrenches will be liable to remove the corners of the nut. On the other hand, excessive leverage, as with a socket set and long tommy bar, is not necessary, and if persisted in can bend the rocker shaft. A normally dimensioned spanner of good quality will do all that is required.

Instead of a rocker-shaft mounted in cylinder head pedestals, some engines have an individual mounting stud for each rocker as shown on Fig. 1.3. The stud carries a spherically-seated bearing

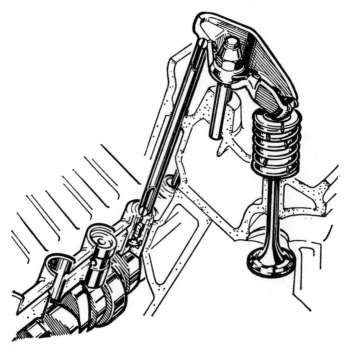

Fig. 1.3. Rockers mounted on individual studs, with adjustment by single nuts, on Ford V-4 engine.

assembly which can be raised or lowered on the stud, by turning the same nut which retains it thereon. The nut may be of the 'stiff' self-locking type containing a Nylon insert, or it sometimes has a separate lock-nut. In such cases there is no adjusting device at the rocker-end and top of the push-rod. The clearance is adjusted by screwing the fulcrum nut referred to, up or down the stud, and measured by feeler gauge as already detailed. Where a self-locking nut is used the job is obviously very simple, but if there is a second lock-nut, care must be taken not to move the setting when locking the main nut. Well-fitting spanners must always be used.

Side-valve engines

Though now obsolete, side-valve engines continue in use, probably the most popular type being the 1172 cc Ford. For many years this unit was fitted with non-adjustable tappets, the correct clearance being obtained at overhaul periods by grinding the bottom of the wide-base valve stem, and otherwise being left well alone. Later engines had a self-locking tappet as shown on Fig. 1.4, requiring the use of only one spanner to adjust.

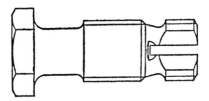

Fig. 1.4. Self-locking tappet used on small Ford side-valve engines.

Other side-valve engines conform generally to the design shown by Fig. 1.5. The tappet end has a hexagon head screwed therein, and provided with a lock-nut. It is necessary to use two specially thin spanners of correct size to tackle the two hexagons, but otherwise the routine to be followed does not differ from that described for o h v engines.

The side-valve engine is inherently somewhat inefficient in terms of power output for a given cubic capacity, and is rather less sensitive to clearance niceties. This is just as well, since the tappet cover plate is usually at the back of the manifolding, and the adjustment points are low down at the side of the engine.

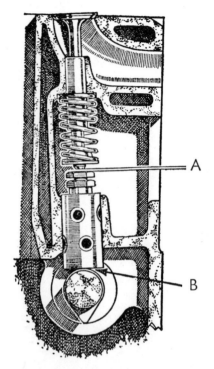

Fig. 1.5. A typical example of a side valve. The clearance is measured at A. B shows the tappet on the back of the cam for maximum clearance.

Overhead camshaft engines

Clearance adjustment on o h c engines is usually done by shims of appropriate thickness, positioned between the end of the valve stem and the inside of the tappet. The clearance is therefore graduated in steps according to the shim thickness, and adjustment requires removal of the camshaft and tappets from above the valves. Once set, however, there is no chance of the clearance adjustment shifting and it is usual for this type of engine to perform well for long periods without the necessity for re-adjustment. The clearance can be checked by using a feeler gauge between the upper face of the tappet and the cam, with the camshaft in position and turned so that the tip of the cam lobe is diametrically opposite to the tappet as shown on Fig. 1.6. The clearance must be determined by the steps

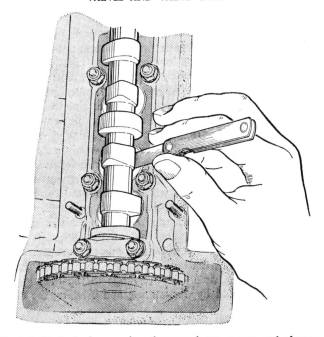

Fig. 1.6. Method of measuring clearance between cam and plunger-tappet on o h c Hillman Imp engine.

of shim thickness as supplied by the manufacturers, and there is thus no point in re-checking with the 'go' and 'not go' feelers as for push-rod engines. This latter method however should be used on o h c engines having mechanism other than plunger tappets and which incorporate a stepless adjustment.

Lubrication of overhead valve gear

So much for valve-clearance adjustment. It will be by now appreciated that a task such as this requires time and patience to do properly, but, once done, will ensure efficient operation for a very long time. While the job is being done there are several other points which should receive attention, though these amount to little more than a visual examination.

A check of the lubrication system can be made while the rocker-box is off. There is no objection to running the engine with the cover removed, so long as the atmosphere is clean, but it is advisable to protect the adjacent coachwork from stray oil splashes. Some valve-rockers have an oil-hole or drilling emerging at one or more points, designed to feed oil on to the valve-stem and push-rod ends at the points where maximum loading occurs. Initially, the oil is fed through the hollow rocker-shaft which it reaches either by way of an external pipe, or oil-ways in the engine castings, or both. The location of the oil holes in the rockers can be ascertained with the engine at rest. It is then a simple matter to watch that oil flows freely from the holes with the engine running at normal idling speed. It should be noted that the flow is quite liberal, in other words, a steady stream, and any marked difference in flow between oil holes in similar locations must be suspect. No flow at all is a serious matter, as it will be appreciated that the engine will still continue to run with little or no lubrication to the rockers, the result being rapid wear of expensive components, not to mention excessive noise and inferior performance. However, assuming that all the oil-feeds deliver as they should, all is well. If the flow from one or more is intermittent or deficient in quantity, the fault must be found. It will usually be nothing more serious than dirt or sludge in the drilling, a fault which can be remedied by dismantling and cleaning the offending components.

In the case of individually-mounted rockers which have no rocker-shaft to act as an oil distributor, the oil feed may be taken via the hollow tappet and push-rod from an initial pressure point surrounding the tappet. It is very simple to carry out a visual check because of the complete isolation of each rocker from the adjacent ones.

Another check which can be carried out with advantage to the owner's peace of mind is the security of the valve-spring retainers. The top valve-spring caps are usually held by a split-cone cotter, which is a snug fit in a groove in the valve stem. Viewed from above, the assembly appears as a pair of half-circles with small gaps in their abutting ends. Points to check are that both half-cotters are flush with each other and approximately so with the cap itself. Excessive wear, which is most unlikely with modern design and materials, renders the cotter liable to pull through the cap orifice, but any tendency for this is easily seen by the above examination.

The top ends of the push-rods should also be examined, the oil film being wiped off to enable this to be done. The cups engaging the rocker ball-ended adjusting screws are often a press-fit in the push-rods, and sometimes they tend to open out the top of the rod, so becoming loose. Any tendency towards this will be obvious.

Many engines are provided with oil deflecting and sealing arrangements under the top valve-spring caps, to prevent an excess of lubricant from passing down the valve stems and into the combustion chambers, a typical arrangement being shown in Fig. 1.7.

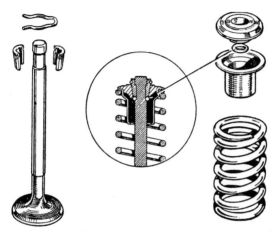

Fig. 1.7. Components of oil-sealing device at top of valve-stem, on B.M.C. engines.

It is not possible to check the condition of such seals from a visual examination with the components in place, but no attention is usually called for between top overhaul periods. Failure is generally accompanied by excessive exhaust smoke and other signs of too much oil in the cylinders.

Oil-tightness

At one time, the replacement of the rocker-box was merely a matter of putting it on and screwing down the holding nuts. Nowadays, it has become a major problem on many engines, due to the use of very light steel pressings for the cover. These may only provide a somewhat sketchy width of flange to form the joint with the cylinder head, and are also liable to distort at the slightest provocation, more particularly if extra force is used on the securing nuts in an effort to ensure oil-tightness.

Distortion of the cover, if it is of the type outlined, must be avoided at all costs. Should it have occurred it is possible to rectify things by placing the cover on an absolutely flat surface and observing what sort of a fit is obtained between the bottom flange and the surface on which it is sitting. Very careful tapping with a light hammer on the flange edge can usually correct matters. When this has been done, the only way to ensure oil-tightness is to use a soft packing between the cover flange and the faced surface of the cylinder head. The packing washer supplied by the makers should be suitable, but if for any reason this is unobtainable, sheet cork or similar composition having a goodly degree of resilience is required. The washer should overlap the flange edge by an adequate margin, to allow for the inevitable spring of the cover when it is tightened down. Liquid jointing compound should be applied to both sides of the washer, which is then laid on the cylinder head in the right position. The cover should then be placed over its studs, and pressed on to the packing, the latter being positioned with the fingers to give as even an overlap all round as possible.

If the cover holding-down nuts are of the finger-tightened type, they must only be dealt with by hand without the aid of spanners or pliers. With hand-tightening, it is unlikely that distortion of the cover need be feared. If, however, hexagon nuts are used, it is all too easy to apply excessive force, and this must be avoided. Sufficient tightening pressure to compress the packing washer is all that should be required. After a few days' running, a little extra tightening may be called for, if the washer and joint has settled down.

Overhead camshaft engines generally have substantial cast camshaft housings, as there is a good deal more oil present inside, and there is also extra noise to be damped as much as possible. Such covers must be fitted with the care already detailed, with all washers in good condition and the nuts tightened evenly. This type of cover is not likely to distort under any conditions, but should nevertheless be fitted with great care, noting that the washers must be oil-tight and the nuts tightened evenly.

If a breather-hose is fitted to the cover, it must be replaced as soon as it shows sign of deterioration due to oil, heat or other causes. Its hose-clips should be fitted with the same care as in the case of the water connections, which are dealt with in Chapter 7.

It may be thought that a lot of space has been given to this question of oil-tightness, but there is little doubt that the rocker-box flange is one of the most frequent 'leakers', and the evidence provided by many suburban drive-ways indicates that there is still far too much unwanted oil coming from somewhere.

The routine in this Chapter should be carried out in accordance with the car manufacturer's servicing schedule, which for push-rod engines is 5,000 to 8,000 miles. Bear in mind that once everything is on the top line, further energies will largely amount to visual examination only, over long periods of running.

Chapter 2

Spark Plugs

*Spark plug welfare—Plug examination—Plug cleaning—
Non-detachable plugs—Replacing in the engine—Sparking devices*

Before going on to detail the next task in our tuning sequence, there is a general piece of advice called for, which applies almost without exception to whatever job is undertaken.

This is—take a good look at what is intended, before starting the job. Many instruction books are somewhat vague in describing dismantling operations, while others are commendably complete. Even in the latter case, however, some owners seem to be incapable of following the written word at all closely, and rely more on visual evidence and maybe blind reasoning. It is no use attacking every nut and bolt in sight and hoping that eventually the item which is the object of the attack will emerge into the light of day. A preliminary survey, and the use of normal reasoning powers, will save a lot of time. On the other hand, the use of brute force such as in the matter of bending things out of the way instead of releasing a further attachment point or two, is also to be ruled out of order, whatever the temptation.

Spark plug welfare

Many years ago a world-famous spark plug manufacturer used the slogan 'Fit and Forget'. Even at a long distance in time, there is some evidence that many motorists still do this quite literally, to the detriment of performance. Ensuring that the plugs are functioning as they should is a simple task that pays dividends in both m p h and m p g, and may well be next on our list.

If maintenance is to be given the serious attention warranted, it is well worthwhile to invest in a spare set of plugs, so that the set removed from the engine can be attended to at leisure without prejudicing performance. The danger is, of course, that the set removed may be put on one side and forgotten, hence the reference to serious attention.

All makers have their own recommendations regarding type and make of plug, while the various plug manufacturers also market types to suit every vehicle. Providing the correct type is used, there is absolutely no harm in experimenting with different makes of plug, and in fact this may well be beneficial, as each individual engine has its own idiosyncrasies.

It is however very important not to use a type or grade of plug which is unsuitable for the engine. A plug which becomes excessively hot, causing pre-ignition, may raise the combustion chamber temperature to an extent which could cause piston collapse, particularly in a modern high-compression engine. Plugs are graded by their makers according to heat range required. It will be seen from Fig. 2.1 that the shape of the centre electrode is similar on both

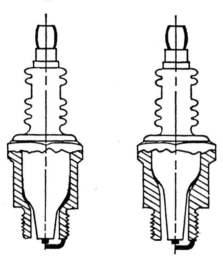

Fig. 2.1. Sections of typical spark plugs, showing different shapes of insulator tip for temperature grading.

plugs, but that the plug on the left has only a short length of insulator at the tip. This type is used on engines having high combustion chamber temperatures, and gives a short heat path to the main body. The plug on the right has a long heat path to provide the opposite effect, and would tend to run at a higher temperature and thus burn off deposits. This plug is therefore suitable for lower combustion chamber temperatures. The general rule is to use a cool plug for a 'hot-stuff' engine, and a hot plug for milder types.

Every 6,000 miles is not too often to have a look at the plugs. When removing them, the washers should not be forgotten should they remain behind in the plug recesses, as frequently occurs when the latter are deeply pocketed. It is a good guide to condition of the engine, to keep a record of the appearance of each plug as it comes out. For instance, the notes might read: No. 1, oily; No. 2, black; No. 3, sooty; No. 4, dark grey. The usefulness of such information will be evident at future examinations, and will be detailed later on.

Elementary maintenance of plugs concerns cleanliness of the exterior, and gas-tightness of the washer. The plugs should never be screwed into their locations in a heavy-handed manner, as this merely makes removal difficult and does not help gas-tightness. Given reasonable consideration on this point, the washers will last a long time and ensure a gas-tight seal. If they are squashed beyond recognition by heavy-handed spanner work, they should be renewed. A washer in good shape, on a plug tightened to the correct degree, will ensure a 100 per cent seal and a high degree of heat dissipation from the plug body.

Dirty plug exteriors enable leakage of current to take place across the insulator. This can be avoided merely by wiping with a cloth. An oil-film on the plug naturally attracts dirt, which is another reason for the avoidance of oil-leaks, as mentioned in Chapter 1.

Plug examination

Examination of the firing end of the plug will convey a lot of useful information. The following is a summary of conditions inside the engine which may be deduced from the plugs' appearance:

If the electrodes are grey in colour, the body end slightly sooty, and the insulator end of a brown tinge, there is nothing much amiss with either the selection of plug or the running mixture.

A light, almost white slatish colour on the end of the body and electrodes indicates an unduly weak mixture; the electrodes may also show signs of burning at the tips. A plug which is of a type insufficiently 'hot-stuff' for the engine may show similar symptoms, but usually affects the running in other ways, and this is fairly easy to diagnose. A carburation defect, such as a small air-leak in part of the manifold system, can, however, easily affect one cylinder only, and is shown up as indicated.

Black deposits on the end of the plug body and electrodes have to be classified into oil and soot. The former is usually rather thick and coarse in grain, and may indicate that too much oil is present in the combustion chamber. This in turn can be caused by defective pistons or rings, too liberal use of upper-cylinder lubricant, or leakage down the valve-stems in o h v engines. Soot, however, is smooth, fine-grained, and usually lightly deposited, being generally the result of too rich a carburetter setting, though, of course, a carburetter defect such as flooding can also be the cause. It is possible, though uncommon, for a mildly sooty deposit to occur even when the mixture is just about correct, if the plug is running a shade too cool for the engine, *i e*, if it is of too 'hot-stuff' a type for that particular duty.

Having noted the condition for future reference, the job is now to restore the plugs to full efficiency. The general construction of plugs varies mainly in regard to whether they are of the detachable or non-detachable type. In the former case the plug can be dismantled for cleaning, while with the latter type, special plant is necessary for thorough removal of carbon.

Detachable-type plugs are seen less frequently nowadays for two main reasons. First is the availability at most garages of very effective cleaning machines; and second, the difficulty of achieving a gas-tight joint without crushing the insulator, in modern conditions of high combustion chamber pressure and temperature. The extra joint and gland-nut also form a break in the heat path from insulator to body, whereas the non-detachable type has its insulator sealed into the body by a special process.

Plug cleaning

When it is desired to dismantle a detachable-type plug, the only really satisfactory method is to grip the plug body in a vice, which must not be screwed up too tightly, otherwise distortion of the body is possible. A close-fitting spanner, preferably of the box type, is then used to unscrew the gland nut. After removal of this, the components of the plug will readily come apart, and care should be taken not to lose any loose washers incorporated in the sealing arrangement between the insulator and body.

Any obvious means may be employed to clean off the carbon from the central electrode, insulator, and interior of the body, the only point to watch being that abrasives must not be used on the insulator, which can be scraped with a knife-blade and given a good wash in petrol, a process which will give an adequate degree of cleanliness. Emery cloth can be used with advantage on both the electrodes and the body interior, petrol being then used to remove every trace of carbon and abrasive before assembling the plug.

When the plug is dismantled it is easy to examine the electrodes or 'points' for signs of burning or pitting. This will not be excessive unless the plug is of an unsuitable type, or has done a very big mileage: 10,000 miles is a satisfactory life for the average inexpensive type of plug, and a new set at this mileage is not to be grudged should the points appear at all doubtful.

When assembling the components of a detachable plug, the main item is to tighten the gland nut firmly but not excessively. This should be done in the vice, in a similar manner to that used for dismantling. The threads on the outside of the plug body should be cleaned out with a knife-blade if there is any dirt or carbon present, as this will forestall any chance of seizure in the cylinder threads, and will also enable the degree of tightness, when replacing the plug, to be gauged to a nicety.

Non-detachable plugs

The recommended method of cleaning non-detachable plugs is by a blast-machine using compressed air plus a mild abrasive. This is, of course, a service station job. The life of the average non-detachable plug is not much over 15,000 miles and the writer must confess that he has never found it necessary to resort to the use of such a machine within this mileage. Removal of the carbon by scraping from the plug end, and careful attention with a small wire brush between the insulator and the plug body will enable all carbon, likely to cause trouble, to be removed without recourse to more complicated methods. A final wash in petrol will complete the job.

If the plug points have been trimmed carefully with emery cloth and a fine file, there should be no difficulty in setting the gap with an ordinary feeler gauge. The usual gap these days is about .025–.030 in., but earlier engines, particulary of fairly high-compression type, used much smaller gaps. Thus, care must be taken in arriving at the best gap, for elderly and vintage cars, as their coils and magnetos were not designed to deal with the large gaps which are now commonplace.

The plug gap is set by careful bending of the side electrode. The central electrode must never be bent, as this will damage the insulator and make the plug useless. Bending of the side point is best done by using a tool-cum-gap gauge as shown on Fig. 2.2, but can also be done by means of thin-nose pliers or by light tapping. The feeler gauge should be a tight fit in the gap, and in fact the feeler can be used as a distance-piece between the points when obtaining the final adjustment by light tapping.

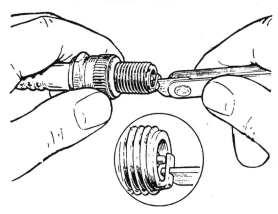

Fig. 2.2. A special tool combined with gauge, can be obtained for 'gapping' plugs, as shown.

Replacing in the engine

The undesirability of over-tightening plugs when replacing in the engine has already been mentioned. Many plug spanners supplied as standard do not encourage proper appreciation of this point, as some plugs enter the cylinder head at a peculiar angle, and in consequence the plug has to be 'aimed' properly before it will engage the threads. A well-fitting box spanner is essential, and must be of sufficient length to clear the surrounding metalwork, at the same time being thin enough to provide clearance if the plug is deeply recessed in the cylinder head. Should the outside of the spanner foul the plug recess it will be almost impossible to enter the plug in the hole, or to tighten it properly.

The most satisfactory form of spanner is a normal box spanner of well-fitting tube type, having an extension tube of the necessary length welded to one end, and a tommy-bar similarly attached T-wise to the other. With a tommy-bar about 4 in. long, sufficient tightness will be imparted without overdoing it, and the one-piece construction will be a boon. The tool costs very little to have made up, and is just as effective as the so-called racing plug spanners which appear to be the only marketed alternative to the standard loose tube spanner and detachable tommy-bar.

There is one final emergency tip. Should the side electrode break off a plug, and no spare is available, forget the no-bending rule in regard to the central electrode, and very carefully bend it sideways until it is just clear of the plug body. It may get you home.

Sparking devices

Apart from the wide of choice of spark plugs manufactured by established and well-known makers, other types appear from time to time for which various intriguing claims may be made. As already mentioned, it is quite possible to damage an engine quite seriously by a wrong choice of conventional plug, and this must clearly be borne in mind in any experiments with unusual designs. Plug makers spend an enormous amount on continuous research and it is logical to assume that they know most of the answers in providing trouble-free firing of the charge. If one wishes to try a device marketed under an intriguing name because of various claims made for it, the best policy is to ask the car manufacturer for his opinion before doing so. It should also be mentioned that many of the ills which these plugs are supposed to cure, may be the result of mechanical wear or neglect of the engine; such faults should be put right before blaming the ignition.

CHAPTER 3

Ignition Equipment

*The contact-breaker mechanism—Modern wiring—The make-and-break—
The automatic advance mechanism—Gap setting—Final assembly—
Timing for efficiency—The distributor position—Road testing—
Optimum setting*

Having now dealt with spark plug welfare, it is logical to follow on with information regarding the routine to be followed to enable the ignition system—the components providing the 'sparks'—to function in accordance with their full capabilities.

Fuel is now available in greater variety and of higher quality than was the case only a few years ago. Regardless of the age of the engine, however, it is important to set the ignition timing to suit the grade and octane rating of the petrol to be used. The aim should be to obtain maximum performance, with the fuel selected accordingly, rather than attempting to economise by running on low-octane petrol with the timing undesirably retarded. Timing the ignition is a fairly exacting task and before detailing the routine to be followed, it is necessary to ensure that the whole of the ignition system is in good condition.

Normally, attention to such items as the contact-breaker and distributor is not required except at infrequent intervals of about 6,000 miles. It is, of course, possible for the system to continue working, with a measure of success, for an almost indefinite mileage but if this is allowed to happen it is likely that a good deal of work and some expense will be called for when the inevitable finally happens. Regular attention, on the other hand, ensures reliability and takes up very little time.

The contact-breaker mechanism

The modern contact-breaker and distributor unit has been developed to a very high degree of reliability and such items as broken springs and faulty condensers, at one time rather commonplace defects, occur most infrequently. The condenser still comes in for more than its share of blame for certain ignition faults, but its misbehaviour should never be taken for granted.

When examining the contact-breaker unit, the most satisfactory preliminary is to remove the distributor cover and its high-tension leads in one piece, rather than attempting to work with the cover slipped on one side where it usually gets in the way. It is necessary to disconnect one end of the high-tension lead between the coil and distributor centre, to enable the cover to be taken clear away, and usually the coil end is easier to detach. It may be secured by a moulded terminal having a milled edge to facilitate unscrewing, as shown on Fig. 3.1. If this is abnormally tight, pliers may be used to

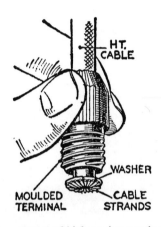

Fig. 3.1. Arrangement of high-tension terminal and cable.

assist the unscrewing effort, but very great care is needed, otherwise the terminal will be cracked. ('Finger-tightness' should always be the rule on these terminals; they will not come loose if properly tightened with the thumb and finger.)

Should the leads to the plugs be enclosed in a tube, or restrained by cleats, it may be simpler to remove the attachments of the tube or cleats, rather than attempting to withdraw the bunch of leads from the restraining influence. The distributor cover is held by two spring clips, and is readily removed after springing these back. In the centre of the cover, inside, will be found a small spring-loaded carbon brush. Care must be taken that this is not lost, but if it is in good condition (and it has an almost indefinite life), there is no need to remove it. Make sure that it is firmly gripped by the end coils of its spring, and that the far end of the spring is lodged securely inside its moulded recess in the cover. It will be realised that the whole of the ignition current passes through this spring, hence the need for secure contact at these two points.

No liquid of any description must be used when cleaning the cover. The outside usually collects a film of oily filth which should be removed with a dry soft cloth. The high-tension cables should have similar attention, and also their enclosing tube, if such is fitted. The anchorage of the cables in the distributor cover can be readily tested by pulling on them, but do not use too much strength, particularly if their attachment takes the form shown on Fig. 3.2

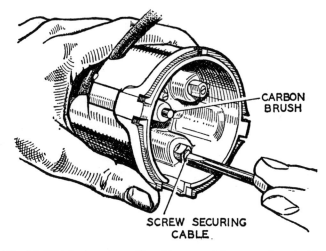

Fig. 3.2. Distributor cover with cables secured by screws from inside.

whereby the cable ends are secured by pointed screws from inside the cover.

It is usually possible to see signs of corrosion, if such are present, and if corrosion is suspected, the terminal should be detached for thorough examination. Moisture not infrequently finds its way down between the lead and the moulded terminal at the distributor end, and, of course, once inside, it cannot escape, but remains to attack the internal copper contacts. A mysterious misfire is sometimes directly traceable to this cause, but rectification of the fault is simple. The defective portion of the cable should be cut off, and a new end formed by cutting back the insulation, leaving clean and bare wires to contact the terminal at the bottom of the recess. The latter should be cleaned also, a small knife-blade being effective in scraping out any dirt or corrosion. After reassembly, a little liquid jointing compound run down between the high-tension lead and its moulded terminal screw will prevent any moisture entering in future.

Some distributors have side-entering leads, which are secured by screws accessible from inside the cover. The screws have points to penetrate the insulation and form a contact.

The leads themselves only become faulty if carelessly handled or made to bend to an unduly small radius, when cracks may appear in the outer insulation. It should be emphasised that when in good condition, the leads will not come to any harm if allowed to come in contact with metal (at normal temperature, of course, not the exhaust pipe), and elaborate methods of ensuring that the leads do not touch metal are quite unnecessary. Obviously, however, they should not be allowed to chafe, but if properly constrained they will not only look neater, but will last longer, and there will be less chance of a lead becoming detached from a plug terminal at awkward moments.

Modern wiring

The modern type of cable used for high-tension leads has an oil-resisting outer sheath with an almost indefinite life. Even when bent sharply, this type of cable is remarkably free from cracking, but it is far better to use plenty of cable length, and give it an easy run with gradual curves; it is recommended that when cables are being renewed, this type is used. The toughness of the covering makes a really sharp knife essential for removal of the insulation when attaching the terminals.

A non-metallic conductor is now coming into use in place of the old-established multi-strand copper cable, and this is an important point when renewal is required. Non-metallic cable has a specific electrical resistance built in to it for the purposes of radio interference suppression. Should replacement be called for, the correct type and length of cable must be obtained from the car servicing agent, and cut to the same length as the one being replaced.

Oil occasionally finds its way inside the distributor. If the outside is kept clean, there should be no infiltration from this source, but too much enthusiasm with the oil-can on the mechanism inside can have a bad effect. The inside of the moulded cover should be cleaned in a similar manner to the outside, any defects such as cracks then being shown up readily. These are unlikely, unless the cover has been roughly handled. The inside contacts may be slightly burned, and the carbonised coating can be removed with a knife-blade, a final polish being given with fine emery cloth. This completes work on the cover, leads and terminals.

Removal of the rotor arm from its shaft is merely a matter of pulling it straight off its boss without any twist. There is danger of breakage if twist is applied, as the rotor has a driving key moulded on its inside, which engages a keyway in the driving boss. The arm must be a good fit; if it is wobbly or sloppy but otherwise in good order, a better fit can sometimes be obtained by building up the inside of the hole with shellac varnish, leaving this to harden thoroughly before fitting the rotor. A second coat can be given, if one is not enough.

After removing and cleaning the rotor, the brass wiper arm should be examined. Its edge may be somewhat burned, but if not excessively pitted, can be trimmed with a fine file to a smooth contour. There is no harm in removing two or three thousandths of an inch of metal, but if this does not suffice, a new rotor should be obtained as the jump of the current to the distributor contacts will be excessive. A final finish with metal polish is not wasted; clean and bright metal goes very well with the passage of electric current.

The make-and-break

Modern types of contact-breaker are somewhat simpler to dismantle than was the case a few years ago, though most of the improvements relate only to details. A 'vintage' type is shown on Fig. 3.3 from which it will be noted that the ignition timing is controlled by hand from a cockpit lever, and that the assembly is of a simple type, the adjustable contact being mounted on the end of a locknutted screw. The type shown on Fig. 3.4 incorporates a centrifugal governor mechanism which varies the timing according to engine speed, while Fig. 3.5 illustrates modern practice, with the timing controlled both by engine speed and induction manifold vacuum. The information following applies generally to all types, but is particularly applicable to that shown on Fig. 3.5.

The low-tension wire should be removed from the terminal on the outside of the distributor unit, and the insulating bush and condenser lead taken off. This will release the end of the flat spring of the moving contact. Earlier types of spring may have a hole at the end instead of a slot, and the terminal bolt will have to be pushed right out to free the spring.

With the end of the spring released, there is no obstacle to with-

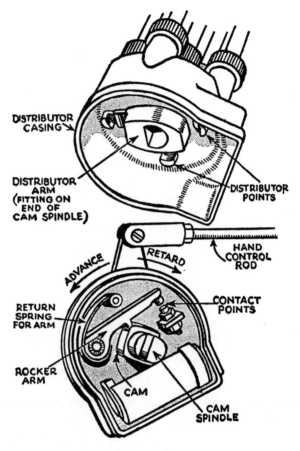

Fig. 3.3. Old type of contact-breaker mechanism, with hand-control of timing.

drawing the contact rocker-arm off its pivot. Under it, on the pivot, will be found a fibre insulating washer, which should also be removed. The fixed contact, the support plate of which is fitted over the same pivot pin, can then be taken off after removing its securing screws, taking care not to lose any of the spring washers.

If at this stage everything else seems to be in a clean condition, there is no need to carry the dismantling operations any further, but if in doubt it is as well to have a look at the automatic advance

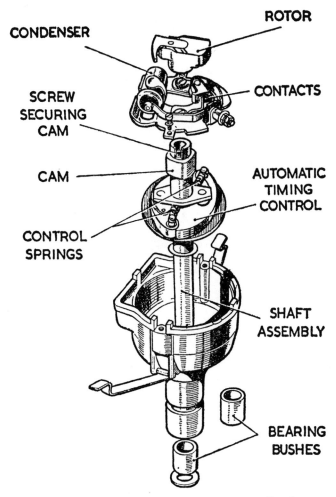

Fig. 3.4. Components of distributor unit with centrifugal-governor control of timing.

mechanism. The latter is exposed by taking off the contact-breaker baseplate, which is held by two screws at its edge. The condenser will come away with the baseplate. In earlier types the condenser is held in a bolted clip, but modern versions have a soldered mounting.

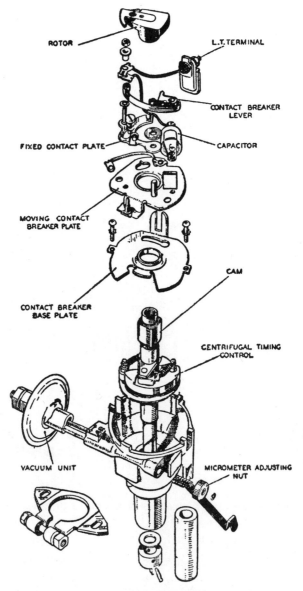

Fig. 3.5. Components of distributor unit with centrifugal and vacuum control of ignition timing.

IGNITION EQUIPMENT

The whole of the parts removed should be cleaned carefully with a dry cloth, after which the contacts should be examined. It will usually be found that there is a pimple in one contact and a corresponding crater or depression in the other, and neglect causes these eventually to assume quite impressive proportions. Modern contacts have plenty of metal to play with, and this is just as well, since restoration of neglected 'points' demands quite a lot of filing. In really bad cases it is probably best to renew the points—both of them. However, if renovation will meet the case, as it will on a well-kept engine, there is no particular rule to be followed, except a steady hand and the use of a smooth file, followed by an emery slip used with delicacy. A perfectly flat surface on the contact

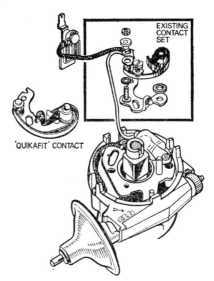

Fig. 3.6. Lucas 'Quikafit' contact assembly compared with the original equivalent parts.

should be aimed at, and no ridges should be left. The contacts can then be assembled temporarily in position, on the baseplate, when it will be simple to ascertain if they meet squarely. A little retouching with the file and emery slip may be necessary, but one or two tests on these lines will see the contacts meeting properly. The spring on the moving contact lasts almost indefinitely, and calls for no attention beyond ensuring that its fixing is secure. The same remark applies to all the other riveted attachments to the pivoting block; failure at these points is most unlikely, but can happen if brute force is used, or if some form of jamming is caused.

The automatic advance mechanism

The automatic timing control is of the centrifugally-governed type, the contact-breaker cam being mounted on a sleeve which is a free fit on the drive shaft and can be rotated in relation thereto by the action of the governor bobweights. The action depends for its accuracy on the correct grading of the restraint springs, which are quite light, so that obviously any undue friction in the mechanism can upset the precision of timing.

There are several movable pivots, and their motion can be followed by moving the cam by hand. A light oil to SAE 20 grade should be applied to all pivots and also to the hole in the top of the cam (there is an oil passage here, and removal of the screw which will be seen is not necessary). Complete lack of oil is, of course, very bad, but at the same time care must be taken not to use too much. One drop to all pivots, and two to the top of the cam, is ample. Any excess always seems to find its way to the contact-breaker points and interior of the distributor cover, where it combines with dust to form a possible leakage path for current as well as helping burning of the points.

Most engines have, in addition to the centrifugally-operated timing control, a further medium of control, by induction-pipe vacuum. This calls for no special comment beyond ensuring that the connections at each end of the vacuum pipe are free from air-leaks, and that the pipe itself is not trapped or kinked as by careless use of spanners on other parts of the engine.

While the distributor is partly dismantled, it is a good idea to ensure that any exterior means of lubrication of the spindle are brought to light. Some older types have a spring-lid oiler on the outside, rather low down. Its position invites neglect on some engines, but if one is found, it should be cleaned and henceforth receive regular attention, so as to ensure that the top bush of the shaft bearing is looked after. Modern types with oil-impregnated bushes do not have this additional lubricator.

Reassembly of the contact-breaker baseplate does not present any particular difficulty, especially with the new 'Quickafit' contacts, shown on Fig. 3.6. The main requirement is care in handling the small securing screws, which should be provided with lockwashers and tightened no more than can be done with a small screwdriver of a size to fit the slots properly. The 'stationary' adjustable contact requires similar treatment in reassembly, and its screws should be tightened lightly until the rocker-arm and moving contact have been replaced. The merest trace of engine oil is applied

IGNITION EQUIPMENT

to the rocker-arm pivot before slipping the fibre block of the rocker-arm over it, remembering also to replace the insulating washer which goes on the pivot under the rocker-arm. The spring can then be slipped into position under the terminal head, and the low-tension wire replaced on the terminal. Now inspect the action of the rocker-arm. It must be quite free, with no trace of stickiness, on the pivot, but without side-shake. The two contacts should meet absolutely squarely, with the points set so that their gap when open, with the fibre heel of the rocker-arm resting on the tip of a cam lobe, is ·014–·016 in. An excellent final finish can be obtained by inserting a very thin carborundum slip between the points and allowing them to close on it. A few strokes of the slip will give a perfect finish and intimate contact. Any dust must then be carefully wiped off.

Gap setting

The older type of contact-breaker shown on Fig. 3.3 had the adjustable contact mounted on the end of a screw providing for infinitely fine adjustment by means of a small spanner. Modern versions rely on a moveable plate carrying the contact, but here again there are different methods of fixing. The plate locates on the rocker-arm pivot as already described, so that the relative position of the two points is maintained. When the plate is secured by two screws, as on Fig. 3.7, the most satisfactory method of adjustment is to tighten these screws only sufficiently to compress their lockwashers

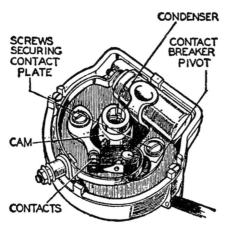

Fig. 3.7. This type of contact-breaker has two screws securing the 'fixed' contact plate.

lightly. Then insert the feeler gauge, with the heel of the fibre block bearing on the tip of a cam lobe to give the fully-open position. Very lightly tap the edge of the plate, using a small flat punch and hammer so as to bring the contacts together, trapping the blade of the feeler between them. If the plate is moved a shade too far, use the punch to move it the other way. In some later types the fixed contact plate is secured by only one screw as shown on Fig. 3.8.

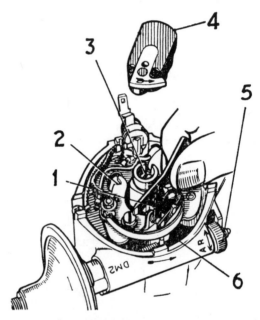

Fig. 3.8. Contact-breaker assembly with single screw fixing for contact plate, and adjusting 'notches'.

1 Gap gauge. 2 Notches for screwdriver. 3 Oilway to cam spindle. 4 Rotor arm. 5 Vernier adjusting nut. 6 Oil hole to governor mechanism.

There is a notch at one end of the plate aligned with a corresponding notch in the base, to allow a screwdriver to be inserted with its blade bridging the two notches. If the single screw is slackened just sufficiently to allow the plate to move, a screwdriver applied in this manner can be used to swing the contact plate about the pivot, with the cam positioned as already stated. After obtaining the correct gap, the single screw must be re-tightened.

Regardless of the precise type of contact-breaker assembly, it should be emphasised that no force must be used on the small screws.

After arriving at the correct setting, and before testing it, the screws should be fully tightened, otherwise the spring pressure of the rocker contact will move the adjustable point and widen the gap. Tightening should be done with a small screwdriver, and after this operation the actuation of the contact-breaker may be examined by testing the gap with the feeler gauge on each separate cam lobe. A variation of more than .001 in. in the gap, when tested in this manner, is undesirable, as it will lead to timing errors between cylinders.

The modern type of polished spring retains its temper almost indefinitely, and breakages are rare. The spring can be tested by applying a small spring-balance to the rocker-arm contact when the latter is in the closed position, and pulling the rocker-arm (opening the gap), by means of the balance. A pull of about 20–24 oz. should be necessary to open the contacts.

The condenser is often unjustly blamed for ignition troubles. In actual fact, it rarely offends, and a defective condenser is nearly always shown up by undue pitting of the contact points and a weak spark at the plugs. If after cleaning, the contact-breaker points start to deteriorate rapidly the condenser must be suspect. Its replacement is not difficult; on some types, a clip fixing is used, while later versions have the body of the condenser soldered to a lug which is a fixture. In this case the faulty part must be unsweated, and the replacement soldered to the lug. Care must be taken to avoid overheating the condenser body when carrying out this operation.

Final assembly

Before replacing the rotor arm, the cam should be given a light smear of engine oil (see Fig. 3.9). Grease may be used if desired, but take care not to overdo things. Oil applied every 1,000 miles or so is to be preferred. The rotor arm should next be replaced, noting that its key engages properly in the keyway of the boss. After a final check to ensure that no stray oil-splashes have been overlooked, the distributor cover can be clipped back in place, and the leads reconnected.

This completes our precision overhaul of the ignition system. Obviously, good performance depends on several other factors as well as the sparks department, but in order to keep things in order, we will assume for the moment that all is as it should be, and finish this chapter by considering how to obtain an ignition timing which will get the best out of the particular fuel used.

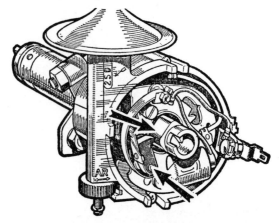

Fig. 3.9. Arrows indicate oiling points for cam and contact pivot.

All makers stipulate an ignition timing which might be termed as standard, this being usually something in the order of—points just breaking with the piston at t d c of the firing stroke on the applicable cylinder. Such a setting provides for satisfactory operation, and no doubt in many cases it is absolutely correct for a particular engine, but nevertheless it is worth while carrying out a methodical check to see whether the performance can be bettered. This, of course, applies with particular force when a change from relatively inferior to a premium grade of fuel is being considered. Also, it is a regrettable fact that some garages seem unable to resist the temptation to tamper with the ignition timing whenever a car is sent in; once the right setting is found, it should be adhered to, the components being marked if necessary so that unauthorised alteration can be spotted at once. The timing is, of course, altered whenever the contact-breaker gap varies; a small gap retards the timing while an over-large one advances it. This will be obvious from a moment's study of the cam movement, remembering that the clearance between the cam and the fibre heel, when the points are closed, is determined by the position of the adjustable contact.

Timing for efficiency

In order to arrive at an ignition timing for maximum performance

IGNITION EQUIPMENT

we have to start by checking the maker's timing, and for this we want one piston at top dead centre of the firing stroke. It is usual to time on No. 1. That is, the cylinder at the front. The piston concerned will be the one which is at t d c with both valves closed. The corresponding piston in a like position on a four-cylinder would, of course, be No. 4, which, however, would be at t d c of the exhaust stroke, the exhaust and inlet valves on the cylinder both being open, though not fully so. Thus, there is no occasion for confusing firing t d c with exhaust t d c on the required cylinder, if the above remarks are considered, but it is nevertheless quite a common time-wasting error. With the rocker-cover off, exposing the valve mechanism, a stiff wire probe inserted through the plug hole of No. 1 cylinder will enable firing t d c to be ascertained roughly, if the engine is slowly rotated. To facilitate matters, all the plugs should be removed to obviate the 'spring' of compression on the other cylinders. The engine can be rotated in the same way as when adjusting the valve clearances, as described in Chapter 1.

As the piston starts on its upward stroke, the inlet valve will be seen to close, the exhaust valve remaining motionless, and this will indicate that the piston is coming up on the compression stroke, *i e*, approaching firing t d c. Care must be taken to ensure that the probe does not come in contact with the cylinder walls in a manner likely to cause scratches. When the piston has reached t d c some degree of accuracy is necessary to determine the final position. It may be that the finger can be substituted for the probe in the plug hole, if the piston comes sufficiently close. Otherwise, a short, stiff rod held against the piston, using the side of the plug hole as a fulcrum, will probably meet the base. The engine should now be gently rocked a few degrees over the t d c point until the absolute mean point is established. It will be appreciated that for quite a distance in the neighbourhood of t d c the piston will appear to have virtually no up-and-down movement at all, and a sensitive touch is essential, particularly in the case of o h v engines having the plug hole at the side; when the hole is right over the piston, the job is, of course, much easier. However, care will enable the mean position to be arrived at, and this can then be accepted as t d c.

This procedure can be omitted in the case of engines which the manufacturers thoughtfully provide with a t d c indicator; this may take the form of marks on the flywheel or a pointer on the timing cover with a notch in the crankshaft pulley. A typical arrangement is shown on Fig. 3.10.

The distributor position

Investigation of the distributor will show the rotor arm to be in line with the distributor contact corresponding to the cylinder under consideration. The contact-breaker points should be just breaking, that is, they should not be fully open to the extent of their maximum working gap, but should be far enough open to allow the insertion of a feeler of one or two 'thou.'. The points may appear to be further open than this, on some engines which are timed for the points to open just before t d c instead of actually at t d c, but if the engine seems to be running reasonably well, no adjustment should be made at this stage. If, however, there is considerable discrepancy, the clamp bolt on the distributor body should be slackened, and the body rotated a shade in the required direction to give the aforementioned opening of the points. It will now be necessary to mark the distributor in some way to indicate the position arrived at. When the distributor is provided with a vernier adjustment by a milled nut, as in most modern types (Fig. 3.8), this

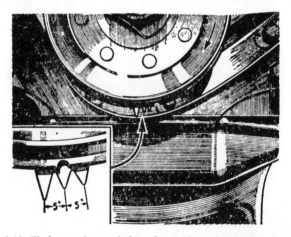

Fig. 3.10. Timing marks consisting of a notch on the crankshaft pulley and three stationary pointers, which indicate t d c, 5° before and 10° before, when the notch is opposite them in turn.

should be set so that equal adjustment is available in both advance and retard directions; this is however only to be used for very fine adjustment, for example to compensate for the unavoidable use of inferior fuel.

For indicating the position of the distributor body relative to its mounting, a good job can be made by rigging up a temporary pointer clamped to a convenient bolt or nut on the engine, and marking the outside of the distributor body with a scriber. The pointer is then bent so as to be in close proximity to the mark and all is ready for obtaining the correct timing on the road.

Road testing

An excellent way of starting road operations would be to fill up with fuel of 'commercial' quality, on which the setting obtained as described would give reasonably knock-free operation in normal medium-compression engines. The ignition would then be advanced until considerable pinking was apparent at speeds up to about 35 mile/h whenever the throttle was opened fairly quickly. The setting thus obtained would not be far off correct for use with premium brands of fuel.

It is, however, asking rather a lot to go to this trouble of changing the tank contents, and in any case a final setting would still be required, obtainable only by actual road test. Further, it is frequently extremely difficult to induce pinking on modern cylinder-head designs, even when the ignition is grossly over-advanced.

A suitable testing venue should therefore be selected, on which the optimum setting can be arrived at using the brand of fuel which is the normal choice. To start operations, we go for that point in the engine speed range at which maximum torque is developed. If the timing is correct at this point, it can be safely assumed that the automatic advance mechanism will look after the timing throughout the power curve. The selected test venue should be a hill which the car will climb on top gear at a speed corresponding to maximum torque. The latter may be stated (in terms of rev/min) in the instruction book, but if not, a rough guide is that maximum torque occurs at about half maximum engine revolutions, which again can be considered as half the maximum speed of the car. In other words, a hill that can be climbed on top gear at 35 mile/h would be an excellent test site for the average family vehicle.

Having selected the hill, the next procedure will be to try a climb at wide throttle in top gear. With the ignition setting as standard, there will probably be no pinking. The speed should be noted, and if this varies somewhat throughout the climb, speeds at easily recognisable points on the hill should be tabulated. Naturally the longer the hill, the better.

This preliminary climb should, of course, be done with the engine at normal working temperature. The ignition should now be advanced slightly, to a degree amounting to only $\frac{1}{16}$ in. or so on the distributor body, the pointer and marking aforementioned enabling this to be judged to a nicety. This is done by slackening the clamp screw and moving the body in the direction opposite to that in which the distributor rotor turns. There is no objection to using the vernier adjustment (if available), for obtaining the final setting, so long as it is not screwed in or out excessively; its purpose is, as stated, to provide a simple way of varying the timing in moderation.

The optimum setting

A test climb should now be made on the revised setting, bearing in mind that performance is the only criterion, and that pinking must not necessarily be expected to occur if the degree of advance is too great. The speed should be carefully noted at the same places on the hill as before. If there is no improvement, make another run to confirm this, and if the same figures are recorded, restore the setting to the original, as no useful purpose will be served by the extra advance. It is, however, extremely likely that there will be some improvement in mile/h. If this happens, advance the setting (with the same routine as before) another $\frac{1}{16}$ in. on the distributor rim, again marking the position. If another run gives further performance improvement, continue the process in steps of $\frac{1}{16}$ in. until no further improvement is recorded. When this point is reached, restore the timing to the setting used for the previous run.

As a final and conclusive check, it is worth while continuing to advance the position for a couple more runs after the 'no improvement' test, when it will be found that the speed actually drops. Do not attempt more than two runs at this setting, however, as it is undesirable to operate on the 'detonating' point for long. Having arrived at the right setting, the clamp screw should be firmly tightened. If vernier adjustment is fitted, there is no harm in seeing if an odd notch or two of movement on the milled nut makes any difference, so long as the original position of the nut is noted in relation to its scale marking on the housing, as shown on Fig. 3.8.

It should also be borne in mind that a deviation in measurement of the contact-breaker gap from its correct setting of 0.014–0.016 in. will affect the timing. Thus it is important to adhere to the correct setting for the points, after maintenance operations, so as to avoid such errors. Obviously, if the maker's instruction book specifies a gap, other than that stated above, it should be adhered to.

Chapter 4

The Induction System and Carburetter

The induction pipe—Freedom from air leaks—The carburetter—The SU principle—Examination of the moving parts—Rectification of faults—Assembling the dashpot—Complications with the jet—Reassembling the jet—The float-chamber assembly—Fuel level adjustment—Spring-return dashpots—SU carburetter adjustment—Starting mixture—Diaphragm-jet type—Control linkage—Special adjustments—Auxiliary starting carburetter—Starter control—The Stromberg CD carburetter—Starting and idling—Normal running—Adjustments—Float level—Constructional points—Fixed-choke carburetters—Solex carburetter—Starting mixture—Adjustment of the Solex carburetter—Selection of part sizes—Examples of jet selection—General maintenance—The Zenith carburetter—Working principle—Zenith adjustment—Slow running—Faults and their causes—Multiple carburetters—Aural check—Reliability

In the previous chapter it was mentioned that when dealing with ignition equipment it is essential that all other items on the engine are operating as they should. This applies in particular to the induction and carburation system, and it must be emphasised that if any doubt exists as to the correct functioning of these components, they should be put right before any attempt is made at precision ignition timing operations.

The induction pipe

As originally conceived, the induction pipe, or manifold, transferred mixture from the carburetter to the engine cylinders, with as equal a distribution as practicable. Nowadays, it also acts, on occasion, as a power take-off for vacuum-operated appliances such as windscreen wipers, brake boosters, automatic timing controls and controlled crankcase ventilation. This point is mentioned to indicate loopholes where unwanted air can enter, and to emphasise that air-leaks can completely upset the whole system.

It will be obvious that all air supplied to the engine should pass through the carburetter for maximum efficiency, apart altogether from the fact that air entering at other places will upset the mixture and render carburation erratic. Atmospheric pressure, however, is a very difficult thing to keep out of any enclosure wherein there is a partial vacuum, and will enter on the least provocation. The first task in checking up on the induction pipe is to ensure that there are no air-leaks.

The flanges might be considered blameless in this respect, as not only do they have a large area of face contact, but are provided with packing washers of special material cut accurately to size. (These packings are almost invariably obtainable for any particular engine, and should be used in preference to home-made packings.) It may be that the very simplicity of the operation of bolting-up the manifold is responsible for the fact that it is by no means unusual for air-leaks to develop at the junction of the flanges with the cylinder or head casting.

The induction and exhaust manifolds are sometimes held by combined clamps which pull up the flanges on both manifolds simultaneously when the nuts are tightened. There is, of course, quite a lot of expansion and contraction at this point during running, and it is usual to find, if one takes the trouble, that the nuts can be tightened appreciably after a few thousand miles. The same remark applies to the flange nuts or bolts between the carburetter and the manifold itself. Unfortunately all these fixings are liable to be inaccessible, but if assembly is carried out correctly in the first place, there will be no need for further attention until the next top overhaul.

Freedom from air-leaks

It will be obvious to start with, that the flanges on the induction manifold should be dead flat and true. The only method of testing the flanges accurately is by testing the faces on a surface plate, but a good check can be made using a rule or other straight-edge, or by placing the manifold, flange faces downwards, on a sheet of plate glass. It is most unlikely that any fault will be found unless there has been some serious trouble such as grossly uneven tightening of nuts. The manifold should be assembled with all flange faces, both on manifold and engine, perfectly clean. A new packing washer of the makers' standard must be used. If for any reason this is unobtainable, one can be made of suitable jointing material obtainable at accessory shops. Thick material must not be used—about the thickness of cartridge paper being all that is necessary.

The washer should be coated on both sides with liquid jointing compound, applied only around the inlet ports if the packing is a combination type taking in both inlet and exhaust ports. No notice should be taken of people who suggest that the use of liquid jointing is unnecessary; use it.

THE INDUCTION SYSTEM AND CARBURETTER

If clamps are used to hold the manifolds, they must be positioned so that they pull up squarely. The clamp nuts must be free on their studs; they sometimes become very stiff, making the degree of tightness difficult to judge, and if this has occurred, penetrating oil should be used to free them, afterwards applying graphite grease to the threads.

The nuts should be tightened up evenly and firmly, and some form of locking washer is advisable behind the nuts. If the manifolds are assembled in this way, there will be no leakage troubles from the flanges. We have, however, already mentioned the existence of other places where it is possible for air to enter, on modern engines. Suction-operated windscreen wipers are a source of trouble in this direction, due to the use of a very long length of flexible tubing, often carried out of sight. This tubing is usually attached at the engine end to a short length of metal piping of small diameter, caulked into the manifold. The pipe may or may not be brazed in, but if the junction is at all doubtful (and some have been examined that were quite literally a push fit), brazing should be resorted to, and a sound joint obtained. The rubber tubing should also be examined frequently and renewed at the first sign of deterioration. This usually takes place at the 'engine' end, due to the influence of heat and oil. No air-leak should, of course, be present in the mechanism of the wiper itself, whether it is operating or not, and if any is suspected, the wiper should be returned to the makers for overhaul.

A rough and ready but effective test is to let the engine idle with the wiper operating, and then with it out of action. There should be no variation of engine speed under these conditions. Then uncouple the tubing altogether from the wiper, at this end, and plug its bore temporarily. Again, there should be no variation in engine speed. If the engine idles more slowly with the tubing plugged, it is an indication of an air-leak in the wiper, the extra air giving a faster idling speed.

Vacuum-operated ignition timing controls are another possible source of unwanted air entry to the induction pipe, but if the pipe unions are kept tight, there should be no trouble. The piping should, however, be occasionally inspected to see that all is in order.

The automatic breather valve for positive controlled crankcase ventilation, which is increasingly being fitted, should be maintained properly, along with its hose connections, to ensure that air does not enter the circuit.

As regards the flange between the carburetter (or carburetters) and the manifold, this should be treated in a similar manner to that described for the junctions with the cylinder ports, that is, the correct jointing washers should be used, with a coating of liquid jointing compound. The nuts or setscrews securing the flange should also be rendered shakeproof in some manner. Where setscrews are employed, threaded into tapped holes in the manifold casting, the best practice is to drill the screw heads across two flats of the hexagon, and wire the heads together. It is quite a simple job to do this sort of work on the occasion of a maintenance session when a few parts are removed, and once done, there is the useful knowledge that the fixings cannot possibly slacken off, and no need for periodical tightness checks, which are probably difficult to do in any case when everything is in position.

Some manifolds have small-bore drain-holes or pipes, to allow surplus petrol to get away if the mixture is allowed to become excessively rich during cold starts. These must be left clear.

It was usual many years ago for all nuts and bolts that called for only occasional removal, to be wired or split-pinned. Nowadays, self-locking or 'stiff' nuts are generally accepted in a variety of positions. When new these are perfectly effective, but the stiffness feature may deteriorate after several removals. If this type of nut gives the impression of undue freedom on the threads, it should therefore be replaced, as it means that the locking medium is not biting sufficiently.

The carburetter

It will probably surprise many motorists if they are informed that the 'life' of a carburetter is about 30,000 miles, but such is the case. This refers, of course, to effective working life. It will go on long after that, but 30,000 is a fair mileage at which it can be assumed that some wear will have developed in the throttle spindle bearings, and that the float needle-valve and seating will not be quite petrol-tight. As a matter of fact, throttle-spindle wear need not occur, over an almost indefinite period, if the spindle bearings are lubricated at reasonable intervals, say every 1,000 miles, with SAE 20 oil. A little of this applied to the spindle where it emerges from the carburetter body will eventually be drawn into the bearing surfaces by the vacuum in the carburetter.

However, this is a point which hardly ever receives attention. The best remedy for wear is to replace the unit with a reconditioned one, which can be done at a reasonable figure. After that, give the spindle bearings their due share of attention; this applies, incidentally, to all wearing parts in the throttle control linkage.

The float chamber needle valve is a different proposition. Wear is quite inevitable, due to the considerable flow of fuel which carries with it minute particles of grit. The needle and seating are usually made so that they can be easily detached and replaced by a matched set of components.

The test for needle valve wear is very simple on cars fitted with electric fuel pumps. With the engine switched on but not running (and thus not drawing petrol) the pump should not 'click'. If it does, it may, of course, indicate a fault in the pump itself, which is a matter to be dealt with later. If the pump is in order, any clicking can only indicate that it is actually pumping and therefore there must be petrol passing through the delivery. If this is suspected to be happening because petrol is being forced past a defective needle valve, a further check can be made. With the air-filter removed and the throttle held fully closed, fuel will be seen issuing from the jet, no doubt very slowly, but sufficient to collect in a puddle at the back of the throttle butterfly. A further symptom which may make the needle valve suspect is when, after a bout of downhill running with the throttle shut, the engine misfires, or fires in a 'lumpy' manner when the throttle is opened at the bottom of the hill. This is, of course, due to the excess fuel aforementioned being suddenly drawn into the engine. On the same occasion, the exhaust will be black in colour for a short period until the richness has cleared.

Carburetter overhaul is a quite straightforward matter if tackled with due care, and in a leisurely manner. The best way of commencing is to remove the instrument complete from the engine and take it to a place where it can be worked on in comfort. We will now describe in detail the points requiring attention on that well-known carburetter, the S U, on the assumption that a somewhat neglected sample is about to be worked on.

The S U principle

The S U carburetter operates on what is known as the constant-vacuum, variable-choke principle. The fuel output from the jet is metered to the requirements of the engine by means of a tapered needle which slides in the jet. The needle is moved in and out of the jet by a piston operated by a vacuum cylinder, commonly referred to as the 'dashpot', this cylinder being in communication with the induction side of the engine. The piston which controls the fuel needle also operates an air-slide which, in effect, varies the choke area simultaneously with metering the fuel supply. Fig. 4.1 shows the principle.

The throttle is of the normal butterfly type, controlling the supply of mixture to the engine. The fuel and air supply are under the independent control of the vacuum cylinder which, of course, de-

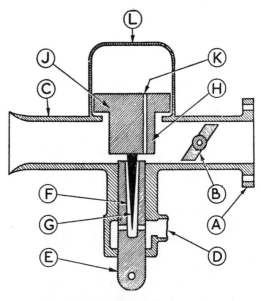

Fig. 4.1. Principle of S U carburetter.

A Mounting flange. B Throttle valve. C Air intake. D Fuel intake. E Jet head. F Jet. G Needle. H Air slide. J Dashpot piston. K Air passage to dashpot. L Dashpot.

pends for its degree of vacuum on the throttle opening and engine speed.

It will be appreciated that all metering of the air and fuel supplies is done by the one moving component comprising the jet needle, air-slide and piston. The jet needle, and the jet in which it slides, are supplied in various sizes by the makers of the carburetter, but should never be changed unless somewhat drastic alterations to operating requirements are necessary. For example, if extreme economy of fuel is essential, it may be possible to gain an improvement in m p g by substitution of a jet needle giving a weaker mixture at certain degrees of throttle opening, but this will only be obtained at the expense of power. The jet and needle combination used by the engine makers is the best for all normal purposes.

Apart from the diaphragm-jet carburetter to be described later, the constructional features do not differ much as between one type

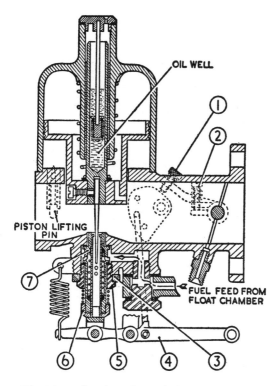

Fig. 4.2. Section through a typical S U carburetter.
1 Fast-idle adjustment. 2 Slow-running adjustment. 3 Washer. 4 Jet control lever. 5 Nut. 6 Jet head. 7 Top guide.

and another. The most obvious difference is that in some cases there is a flexible Nylon pipe between the float chamber and the jet assembly, while other types have the fuel passage drilled in the float chamber attachment. The arrangement is clear from Figs. 4.2 and 4.3.

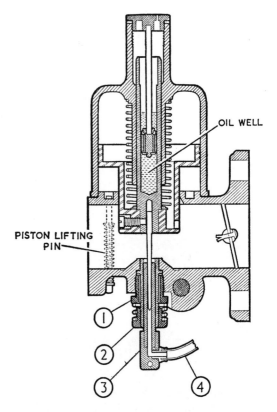

Fig. 4.3. S U carburetter main body, with flexible fuel pipe to jet.
1 Gland nut. 2 Jet adjusting nut. 3 Jet assembly. 4 Flexible fuel pipe from float chamber.

Examination of the moving parts

The operating piston and the chamber, or dashpot, in which it moves, can be removed from the body of the carburetter by taking out the screws holding the dashpot, taking care that the piston is at the bottom of its travel and the needle right down in the jet. As the dashpot is located tightly on a turned register on the carburetter body, it may be stuck. It is most essential that the dashpot is not tilted sideways when removing, as this may result in the needle being bent. The best way of 'unsticking' it is to apply a little penetrating

THE INDUCTION SYSTEM AND CARBURETTER

oil around its base, and then try to twist the dashpot through a few degrees either way, taking care to keep it vertical while doing so. It will be found that once the joint has been broken, the dashpot can be lifted off quite easily. The piston together with its integral air-slide and jet needle should then be lifted out.

With these components on the bench, the dashpot should first be examined. It has a cap-nut at the top of the piston-guide, and this should be removed. Attached to this cap-nut is a thin steel rod carrying a small-diameter auxiliary piston which locates inside the piston rod, and runs in oil. The purpose of this assembly (which is withdrawn quite easily along with the cap-nut), is to form a retarding device to prevent sudden movement of the piston when the throttle is 'jumped on'. This retardation naturally causes a momentarily enriched mixture to issue from the jet, due to the lag in the opening of the air-slide, thus aiding acceleration.

The dashpot can next be examined internally. It will probably be somewhat dirty inside, and should be cleaned with petrol. Next, the jet needle should be removed from the piston. It is secured by a set-screw from the side. This screw must be tight, and a well-fitting

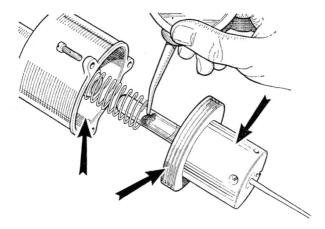

Fig. 4.4. Surfaces shown arrowed must be thoroughly cleaned, and a little oil applied to the piston rod only.

screwdriver is essential both in removing and replacing. The next item is to try the fit of the piston in the dashpot. It should move in and out, for the full length of its travel, absolutely freely, bearing in mind that the aforementioned retarding device, if fitted, has been removed. The piston should also spin easily at any point in its travel —in other words, both piston and dashpot interior should be truly circular. If the piston is spun and reciprocated at the same time, this is a conclusive test for free movement. The periphery of the piston, however, may be dirty. It has grooves turned therein, which allow dirt to settle out of harm's way, and these must be cleaned out. Also, the piston rod and its guide should be quite free from grit, and not oiled while testing. Fig. 4.4 will help to clarify these points.

Rectification of faults

If the carburetter has been neglected, it is extremely unlikely that simple cleaning as mentioned above will result in an absolutely free-moving piston. It is almost sure to bind, however slightly, at some spot. If this happens, clean both piston and dashpot interior with metal polish until they are bright. While doing this it will not do any harm if the same treatment is used on the whole of the air-slide, which accumulates a film of carbon over long periods. After polishing, wash the parts in petrol and try the fit again. If there is still some binding, its exact point of contact must be ascertained. This can only be done by careful feeling, trying the piston in various positions, and holding it sideways against the clearance in the rod and guide. If a high spot is found in the dashpot wall, very careful application of emery cloth will rectify this. If the piston is at fault the same treatment will cure, when applied to the piston periphery. After using any abrasives wash the parts in petrol.

It will be obvious that the effectiveness of the piston and dashpot assembly in operating the carburetter depends on the closeness of fit of the piston in the dashpot. Therefore, metal must not be removed haphazardly. On the other hand, a running fit is equally necessary. The ideal can be arrived at, with care, by the action detailed above.

The jet needle can next be examined. If it is perfectly straight, there will normally be no signs of wear on its surface, as would be caused by contact with the jet. It should not be replaced in the air-slide just yet, as there is a further test to be carried out on the dashpot.

The carburetter body is next attended to. On the type having a Nylon tube to the jet, the float chamber is attached by a bolt. The tube must be detached at the float-chamber end, as this has a gland nut for the purpose, after which removal of the holding bolt allows

THE INDUCTION SYSTEM AND CARBURETTER 47

the float chamber to come away. On other types where the clamp bolt forms the fuel union, take care not to lose the sealing washers when undoing the bolt. The float chamber is now put on one side for separate attention, and the carburetter body thoroughly cleaned. Particular attention should be paid to the register of the dashpot rim, which must be free from carbon or dirt, otherwise merely tightening down the dashpot by its screws may cause distortion sufficient to make the piston foul the interior again.

The various links which actuate the jet (via the mixture control) can be removed; this is a simple operation and makes cleaning much easier. The jet should then be pulled down to the limit of its normal travel, and the outside of it examined. This should not show any ridging, and providing there was no fuel leakage before dismantling, there is no reason to dismantle the jet assembly any further. A smear of petroleum jelly on the outside of the jet will help it to slide smoothly, and reduce the effort needed on the mixture control.

Assembling the dashpot

Before refitting the jet needle back into the air-slide, the dashpot and piston assembly should be replaced on the carburetter body and tightened down, taking care to replace the dashpot the same way round as it came off. It will be noted that the air-slide has a groove down one side, and this registers in a key fitted inside the mixing chamber body. The object of this is to prevent the air-slide and piston turning. If this happened, the air passage into the dashpot, which must obviously face the 'vacuum' side of the carburetter (*i e*, the engine side), would move out of place and the carburetter would not operate.

The air-slide must therefore be registered, with its groove engaging the key, but this is not difficult. After this, the dashpot can be fitted over the piston guide. The object of this fitting-up is to ensure that no distortion takes place when the dashpot is tightened, and after this assembly the action must be tested again. The piston must fall quite freely with no binding anywhere.

If it passes this test, well and good. If not, it must be examined and dealt with once more in the manner already described, until free movement is obtained. Assuming that all is well, the assembly should be removed, and the jet needle refitted carefully into the air-slide. The shoulder of the needle must be flush with the edge of its hole, and the retaining screw tightened firmly (see Fig. 4.5). It is a useful idea, before fitting the needle, to make a note of its number, which is stamped on the shouldered end. This will serve as a check in case of any argument as to whether the correct needles are in use for any particular conditions of fuel or weather.

Having refitted the needle, the dashpot can again be re-assembled. This operation will naturally require more care now that the needle is in position. The end of the needle must be fed carefully into the jet, simultaneously with engaging the groove in the air-slide with its key in the body. The operation is not difficult so long as the parts are kept lined up properly; it is rather easy to get the needle out of

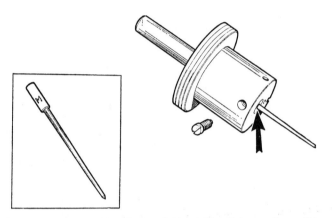

Fig. 4.5. Needle shoulder is fitted flush with piston as shown. Inset shows position of needle identification mark.

the vertical, but this is not so likely to happen with the job on the bench. The needle and air-slide may be allowed to drop to their lowermost position, after which the dashpot can be put over the piston guide and tightened down for keeps.

The oil plunger-retarder inside the piston guide spindle must have the bore of the latter topped up with oil of SAE 20 grade before refitting the retarder plunger. Do not put in too much oil; sufficient to just fill the bore of the guide spindle is ample, and when the plunger is replaced the surplus oil will serve to lubricate the piston guide itself.

Complications with the jet

Before going on to describe work on the float chamber, it will be as well to consider the action to be taken, if when the dashpot assembly is put back complete with jet needle, the latter is found to be binding in the jet. Other troubles warranting further attention would, of course, be a persistent petrol drip at this point, or ridging on the outside of the jet barrel, causing sticking when the mixture control is operated.

In the first case, providing the needle is not bent, the jet must be out of centre. To get it right, it is necessary to centre it, using the needle itself as a pilot. The jet adjusting hexagon—the lowest one—must be tightened up as far as it will go against its spring. Do not use force when tightening this nut, and push the jet head up against it so that the jet is as high as it will go.

Next slacken off the hexagonal locking screw, which is the larger one immediately above the jet adjusting nut referred to. The locking screw need only be loosened a couple of turns or so, and after doing this it will be found that the jet bush, which is encircled by the locking screw (and on to which the jet adjusting nut actually screws), can be moved slightly sideways. With the top cap and damper piston removed from the dashpot, press down lightly with a pencil on the piston rod top, so that the piston is firmly held down, as shown on Fig. 4.6. The needle is now as far down in the jet as it can possibly go. Tighten the hexagonal locking screw, and try the action of the piston for free lifting and falling; on most carburetters a testing pin is fitted for this purpose as shown on Fig. 4.7. If the piston is still binding, slacken the hexagon again, and try moving the jet slightly sideways. A position will eventually be found at which the piston and air-slide, etc., will move quite freely, indicating that the jet is central. Particular attention should be paid to the free movement near the lower limit of the needle's travel, as obviously the more length of needle there is in the jet, the more liability to fouling. When the free position has been found, tighten the locking screw moderately, and test again to make sure that the central position has not been lost. Frequent testing and tightening will eventually get everything correct. Finally tighten up the locking screw firmly, taking care that the Nylon tube to the float chamber, in the case of carburetters so fitted, is correctly positioned to give an easy run-in.

Excessive wear or ridging on the outside of the jet, or fuel leakage from the base of the jet, can only be attended to by a complete dismantling of the jet assembly. This is not a difficult procedure, but it is essential that the parts are removed and reassembled in the correct order.

A jet that is badly ridged externally must be replaced with a new one. Leakage of fuel can be remedied by renewal of the various

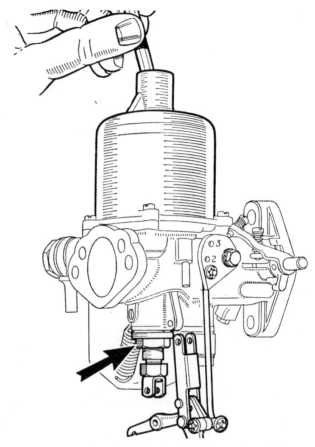

Fig. 4.6. Piston is pushed firmly down as shown, to ensure full containment in jet bore. Arrow indicates locking screw for jet assembly.

packing washers. However, if the jet is being replaced it is best to renew its packings at the same time.

To remove the assembly, the linkage controlling the jet must be removed first. On carburetters having a Nylon feed tube (Fig. 4.3), the large hexagon screw is then unscrewed, which releases the jet holder and gland nut. These can be carefully extracted by tapping, if stuck. On other types the construction is more elaborate, as shown on Fig. 4.8, as the jet holder or bearing is in two halves. To remove, the bottom jet screw, with large hexagon, is unscrewed together with its metal washer and packing washer. This will release the bottom half of the jet bearing, which can be eased out with the jet inside it. The top half of the bearing may remain in the carburetter body and need not be disturbed, but it is necessary for the metal washer and packing washer to be extracted. After the jet has been withdrawn,

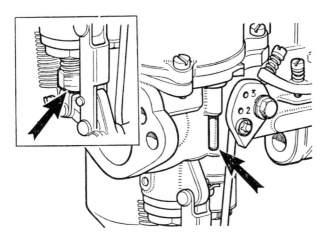

Fig. 4.7. Arrows indicate (*left*) jet adjusting nut (*right*) piston lifting pin used for testing free movement.

the long coil spring and washers can be removed; it will be seen that there are two sets of washer combinations; when in position, these two sets of washers are separated by the coil spring.

Reassembling the jet

Before reassembling the items, everything must be perfectly clean.

If the original jet is to be replaced, its outside should be polished bright with fine emery cloth, to remove any microscopic ridges. New gland and packing washers should be obtained. These are available as a set, at accessory dealers, and no attempt should be made to economise by replacing with other materials.

Refitting the jet with Nylon feed tube is perfectly straightforward and reference to Fig. 4.3 will make all clear. In the case of the two-part jet bearing type, the procedure is a little more complicated, but should not be difficult so long as the right sequence is followed.

With the top half of the jet bearing in position in the carburetter

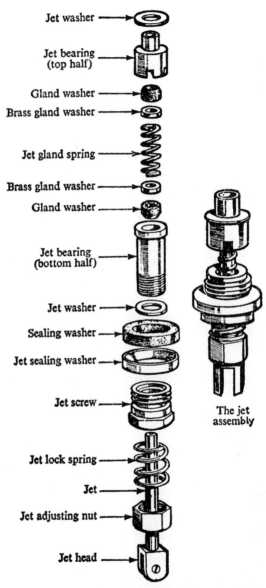

Fig. 4.8. Components of jet assembly, two-bearing type. Follow carefully when re-assembling.

body, the parts for this should be assembled in the following order: a gland washer, followed by a small brass washer. These must be pushed up inside the bearing, and a good tool for this is the spring removed from the jet inside. The bottom half of the jet bearing is now ready for attention. Its main hexagonal fixing screw, with a washer above it, should be slid on. Then the jet adjusting screw complete with its spring and the jet itself can be placed in position, the jet going, of course, inside its bearing. The second gland washer should now be pushed down between the jet and the bearing. The small brass washer goes down next. Again, these items can be pushed home with the jet spring, but this can now be left in position between the jet and bearing.

It will be clear that the spring actually exerts pressure from both its ends (when all is in position) against the two small brass washers, which in turn bear on the two gland washers to form a wedging seal against the outside of the jet, while allowing the jet to slide quite freely.

The main fixing hexagon is encircled by a bevelled brass washer of large diameter which mates with a 'Langite' packing to seal against the carburetter body. Leakage from this point is rare, but if it is thought advisable, as for example when the packing has been damaged during removal, the items can be renewed before replacing the hexagon on the bearing.

After assembly, regardless of type, the jet should be pulled down to its fullest extent and a little petroleum jelly smeared on its exterior. Also, run some light oil through the fuel passages until it emerges from the jet top. There will be no more trouble from worn jets, fuel leakage, or mixture controls which are difficult to operate.

The float chamber assembly

The float chamber may have a fuel filter housed in the inlet union. Removal of the fuel pipe banjo gives access to this filter, which comprises a thimble of gauze, backed up by a light spring. Removal of the banjo usually causes the filter to emerge from its recess, under persuasion of the spring. If it sticks, it can be prised out, but on doing so take care not to puncture the gauze. Never run the engine without this filter, and if it is damaged, fit a replacement. (See Fig. 4.9.)

The float chamber lid may be secured by a central nut, or by a pair of screws at the periphery. There is a packing washer around the lid, which must be in good condition. The lid contains the needle valve mechanism and float lever. The float may be integral with the lever, or it may be of the barrel type and separate. In the latter case it can be dropped out by inverting the chamber, as it slides on a

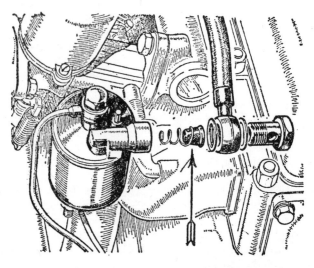

Fig. 4.9. Filter combined with fuel supply union, used on some S U types.

central rod on to which the nut securing the lid is screwed. This rod must be firmly locked in the chamber and, if it is a trifle loose, it should be tightened by locking a pair of nuts on the top threads, and tightening it firmly into its tapped hole by means of these. The whole of the needle valve mechanism is carried on the underside of the lid, and comprises a short needle and seating for same, and a toggle pivoted on a straight spindle, this spindle being located in cast lugs.

Removal of the spindle, which is not retained in any way (except by the walls of the chamber when in position), will enable the toggle and needle to come away readily. Examination of the latter will show whether renewal is advisable. A considerable amount of wear is inevitable after about 25,000 miles even with reasonably clean fuel and assuming the filters are in good order. Wear shows up visually by the presence of a ridge round the chamfered tip of the needle which gives, in effect, two angles of chamfer. The brass needle seating wears, to some extent, to the same shape, but an absolutely fuel-tight seal cannot be obtained with the parts in this condition. If there have been no previous symptoms of fuel leakage past the needle valve, such as those already described, and examination does not reveal undue wear, there is no point in removing the needle seating. If in any doubt, however, the seating should be unscrewed, and both it and the needle renewed. Never renew one part without the other.

The brass needle seating has a hexagon top which can be attacked with a small spanner, and should unscrew quite easily. When refitting the new part do not use undue force, as the threads are small.

Fuel level adjustment

Having made sure that all parts are clean, the toggle lever can be refitted. In the case of separate barrel-type floats, it will be evident that the fuel level is governed by the curvature of the toggle at the point where the toggle-arm bifurcates. As a first check, the float chamber lid should be placed upside-down on a level surface so that the toggle rests by gravity on the float needle, which should

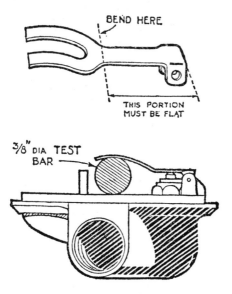

Fig. 4.10. Correct position of float toggle with test bar in position.

be fully home in its seating. In this position it should be possible to place a round rod of $\frac{3}{8}$ to $\frac{7}{16}$ in. diameter underneath the toggle. and resting on the machined circular lip of the float chamber lid as shown on Fig. 4.10. On carburetters where the float is fixed to the toggle lever (Fig. 4.11), the procedure is to bend the lever carefully to obtain the dimension (shown on the diagram) of 0.125 in. between the float edge and the under-face of the lid.

It should be noted that although reasonable accuracy is required, the float level is not highly critical, and the toggle lever should not be bent unnecessarily.

When refitting the float chamber to the carburetter, any doubt-

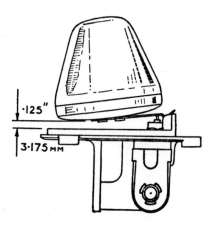

Fig. 4.11. The gap shown must be present to ensure correct fuel level on the type of float shown.

ful washers should be renewed. Do not be too enthusiastic in tightening the large bolt securing the chamber to the carburetter body. It needs to be tight, of course, but remember the alloy threads. The same remarks apply to refitting the fuel-pipe union in the float-chamber lid.

A point to bear in mind on carburetters of the semi-downdraught type, which have the carburetter body at an angle and the float chamber vertically disposed thereto, is that the aforementioned bolt must not be finally tightened until the carburetter is in position on the engine. By this means, the float chamber can be brought to the vertical by trial and error and then finally tightened. When levelling the chamber in this manner, watch that it is not fouling the throttle-stop, as can happen if it is moved too far round.

It may be found on final assembly that when fuel is admitted, a slow leakage persists at the inlet union to the chamber. Do not be tempted to tighten the banjo fitting more than has already been done. Dismantle the assembly, and if all seems perfectly clean and the washers are good, apply a very thin coating of jointing compound to the washers on both their sides, and reassemble.

Spring-return dashpots

S U carburetters of the fully downdraught type have the dashpot and piston assembly horizontal. The piston and air slide-needle combination cannot therefore be gravity controlled, and instead, a large-diameter coil spring is fitted inside the dashpot and around the centre guide. This spring is carefully graded to the carburetter, and must be replaced, if ever necessary, by one of the correct type.

The fully downdraught carburetter is not common nowadays, but the great majority of semi-downdraught or horizontal types are also fitted with springs to augment gravity, and the same comment applies.

After coupling up the fuel supply and pumping-up, the fuel level can be finally checked by lifting up the air-slide (using a stiff wire or cycle spoke) and noting if fuel issues from the top of the jet. If it does not, all is well. A level about $\frac{1}{16}$ in. below the top is about right, but so long as it does not overflow, that is good enough. If it does emerge, do not assume that the level is faulty straight away. Start the engine, and when it is running at say twice idling revs, give the float 'tickler' one or two depressions. This will ensure that any particles of grit, which might have been causing leakage past the needle, are washed away. Then examine the jet again. If the level is evidently too high, it can be corrected by careful bending of the toggle as already described.

S U carburetter adjustment

The slow-running adjustment is under the control of a throttle-stop screw. The hexagon nut at the bottom of the carburetter also regulates the height of the main jet. The positions are shown on Fig. 4.12. The latter nut controls the strength of the mixture, and enables an absolutely regular idling beat to be optained. To adjust the idling, therefore, the throttle-stop is turned as already described, to arrive at the idling speed. The jet nut is then moved with a suitable spanner, but only one 'flat' or one-sixth of a turn, at a time. It is screwed upwards to weaken the mixture, and downwards to enrich it. The effect of the former movement will probably be found to speed up the engine, but overdoing it will cause the engine to 'die'. When this point is reached, turn the nut downwards to the point at which the engine is firing regularly and evenly at the desired slow-running speed. Then turn it one more 'flat' downwards from this point, and that should be just about right. If the nut is turned too far, the exhaust will show a tendency to black smoke, and the beat will become thumpy, with possible misfiring. This is a sign of too

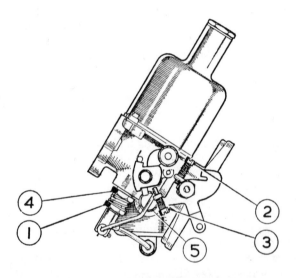

Fig. 4.12. Position of adjusting screws.
1 Jet adjusting nut. 2 Throttle (slow-running) adjustment. 3 Fast-idle adjustment. 4 Jet locking nut. 5 Float chamber fixing bolt.

rich a mixture, therefore the nut should be turned upwards until these symptoms disappear. It should be emphasised that the variation from too rich to too weak and vice versa will hardly amount to more than a complete turn of the nut, so that any movement should only be done a little at a time.

Starting mixture

The so-called 'choke' control, which pulls down the main jet for mixture enrichment, has the great advantage of being progressive. That is, for starting, the jet can be pulled down to its lowest position, allowing the maximum quantity of fuel to pass. As the engine warms up, the control can be gradually returned to normal but throughout, the engine can be kept running smoothly. The only requirement is some skill on the part of the driver in judging the

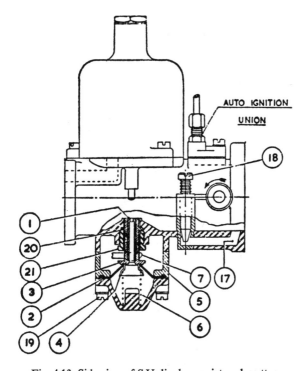

Fig. 4.13. Side view of S U diaphragm-jet carburetter.

1 Jet. 2 Diaphragm. 3 Jet cup. 4 Jet return spring cup. 5 Diaphragm casing. 6 Jet return spring. 7 Jet actuating lever. 17 Mixture passage. 18 'Slow-run' valve. 19 Float-chamber securing screw. 20 Jet bearing. 21 Jet screw.

correctness of the mixture, in relation to the engine temperature; given this, a graduated choke control, providing several positions, is most useful.

However, in some cases the control merely provides a starting position in which it can be 'locked' outwards; when released it springs back and the jet assumes its normal running position. In these cases the engine has to be run for a short time until it is sufficiently warm to be driven with the control released. This places less reliance on the skill of the driver, but also prevents the engine from being over-driven when cold. On some types again, the jet is linked to the throttle spindle in such a manner that operation of the choke control, in addition to pulling down the jet, slightly opens the throttle. This gives a fast idling position which helps quick warming-up and to some extent prevents the car being driven on an unduly rich mixture. Even so, however, it is desirable to return the jet to normal as soon as this is practicable.

Diaphragm-jet type

A development of the S U carburetter described above is known as the diaphragm-jet type, and this is fitted on some cars using

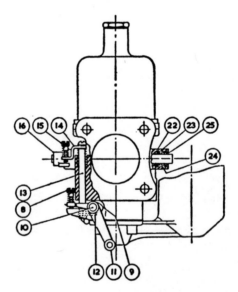

Fig. 4.14. End view of S U diaphragm-jet carburetter.
8 Jet adjusting screw. 9 Cam control lever. 10 Jet lever spindle. 11 Control lever. 12 Cam-shoe. 13 Push-rod. 14 Top plate. 15 Throttle adjusting screw. 16 Throttle stop lever. 22 Cork gland. 23 Retaining screw. 24 Spring. 25 Seal.

choke sizes in the 1.5 to 2-inch bore range. The principle of operation is unchanged, except in regard to the arrangement for supplying the slow-running mixture. Mechanically, the construction is a little different from the type already described, as maintenance of a fuel-seal at the slidable jet is no longer dependent on gland washers. The bottom of the jet is completely isolated from 'outside' by means of a flexible diaphragm (hence the name given to the type) which ensures no leakage of fuel, but at the same time allows the jet to move as required. The diaphragm is firmly clamped between the main casting on its upper surface, and an enclosing cup, which also houses the jet return spring, on its lower face. The construction is clearly shown on Figs. 4.13 and 4.14.

Control linkage

The linkage for controlling the jet is also contained in the main casting above the diaphragm, and comprises a cam movement, giving an extremely light and sensitive action to the dashboard control. An external link couples the jet control to the throttle-stop screw, this link being adjustable in order that the 'fast idling' speed can be regulated in conjunction with the degree of enrichment by pulling down the jet. For slow running, as opposed to fast idling, the mixture in this carburetter is supplied by way of a by-pass instead of through the main throttle valve, though it is still obtained from the main jet The object of the by-pass, which is controlled by an adjustable slow-running screw, is to allow of a rather more sensitive adjustment than can usually be got from the main throttle valve, which is a desirable feature for more exclusive luxury cars. It will be appreciated also that as the idling setting with this by-pass arrangement is quite independent of the main throttle, there is no possibility of it being upset by conditions in the throttle linkage such as maladjustment or lack of lubrication.

Further refinements on the diaphragm-type carburetter include an 'economiser' connection from the through-way to the top of the float chamber, made by a small-bore pipe. The object of this is to vary slightly the 'head' pressure on the fuel when idling, so preventing any chance of an excess of petrol. Another excellent feature is the provision of cork spring-loaded glands on the throttle spindle which eradicate air-leaks at this point throughout the life of the carburetter.

Special adjustments

The jet is centred on this type in much the same way as already described. It is held right 'up' by hand, with its adjusting screw slackened to allow the jet cup to contact the jet bearing, but with a clearance between the adjusting screw and its abutment. Keep the diaphragm in the same position radially (in relation to the carburetter body and jet casing) throughout, it being noted that the jet bore is not necessarily concentric with its outside diameter and therefore it must not be rotated, since this might put it out of centre. If the parts are suitably marked with a pencil there should be no trouble.

The jet is adjusted by a screw instead of the familiar hexagon, but otherwise the principle is unaltered. Slow running is of course adjusted by the separate adjustment screw already referred to.

There are no gland washers to leak, and the only fault in this direction could be caused by an imperfect diaphragm seal. Its outer edge is trapped between the two parts of the flanged joint, and tightening the flange screws will usually cure any leak here. The inner edge, however, is specially fitted to the jet base, and trouble is best remedied by fitting a new jet assembly supplied integral with the diaphragm.

Auxiliary starting carburetter

The hand-operated jet control is almost universal when the S U carburetter is used singly, while even when twin instruments are fitted the linkage required to couple the two jets to the main control is not unduly complicated. In some cases, however (and particularly when more than two carburetters are fitted), the provision for starting makes use of an auxiliary starting carburetter. In this case, the jets in the main carburetters are, once set, left alone, and are not provided with any means for 'sliding' them. The stop-screw for running mixture adjustment has a knurled instead of a hexagon head in some cases, but is otherwise operated just as already detailed. The auxiliary carburetter is arranged for feeding the necessary rich mixture into the induction pipe independently of the main carburetter chokes, and thus it is necessary for the manifold to incorporate a longitudinal duct interconnecting all the in-induction 'stubs'. In fact, such a duct would normally be used in any case, as a balancing pipe for a multiple installation.

The auxiliary unit consists of a housing containing a jet and needle, and is attached to one of the main carburetters at the normal float chamber attachment point, so that it is fed with fuel from the same float chamber. Air enters the housing via a separate inlet, and passes over the top of the jet to an annular chamber surrounding the needle. From this chamber a duct leads to a solenoid-operated valve, and thence to the engine induction pipe.

The jet needle is spring loaded in a manner tending to draw it out of the jet, but is provided with a small disc or piston which is acted upon in the reverse sense by the depression in the annular chamber. Thus, in conditions of high vacuum in the manifold (and thus in the needle chamber) the needle will tend to enter the jet and reduce the flow of fuel to maintain the correct mixture. The degree of enrichment provided by the auxiliary starting carburetter is thus graded to throttle opening, in much the same way as obtains in the main carburetters.

Starter control

The device is entirely automatic, and is brought into action electrically. As mentioned, a solenoid-operated valve is located between the mixture chamber and the induction-pipe duct. This valve is a simple disc-type, closed by a light spring, and opened by energising the solenoid coil, which draws up an armature and thereby lifts the valve. The solenoid can be switched-in by a thermostatic switch, in which case the latter is usually housed in the engine water jacket, and completes the electrical circuit at water temperatures below about 30 to 35 degrees C. When the ignition is switched on from cold, the solenoid will automatically come into action, and the starting carburetter will continue to give the additional enrichment to the mixture supplied by the main carburetters until the engine has warmed up to the desired degree. When starting hot, of course, the device will remain out of circuit, as the thermostat switch will not operate.

In some cases, a hand-operated switch is preferred, usually combined with a warning light on the instrument panel. It is possible that slightly better fuel consumption will thus be obtained, as the device can be switched off as soon as the driver 'feels' that the engine can pull adequately on the main carburetters only. However this involves another control, to be used with some degree of intelligence, whereas the thermostat switch removes this responsibility.

Adjustment of the starting carburetter merely concerns the stop which limits the needle movement, the size of needle and dimensions of its suction disc, and the spring loading. Except for the first, all these are determined by the engine makers and should need no attention. The position of the needle stop determines the amount of mixture enrichment provided for starting and idling with the auxiliary carburetter in action. To obtain the correct setting the engine should be run until it is at normal operating temperature. Then energise the solenoid to operate the starting carburetter, and turn the stop screw until the exhaust gas becomes black, indicating a very rich mixture. Do not overdo it to the extent that the engine beat becomes very irregular. The mixture should then have the correct degree of richness to give a quick start from cold and good warming-up performance.

Where the solenoid is normally under thermostat control, it will of course be necessary to energise it direct from the battery, while carrying out this adjustment.

The S U variable-choke carburetter has been very well-established for many years, and is fitted to a wide range of cars. However, the Stromberg type C D (constant-depression) carburetter, which operates on exactly the same principle, is becoming very popular, and this will now be described.

The Stromberg C D carburetter

Instead of a floating piston, controlling the air-slide and jet needle, this carburetter employs a flexible diaphragm, as shown on Fig. 4.15. It will also be seen that the float chamber is concentric with the main body surrounding the jet assembly. There are three main aluminium castings comprising the main body, float chamber, and suction chamber cover. Alloy castings are also used for the air-valve body and jet housing. For starting, there is a device interconnected with the throttle, providing a fast-idle position for warming up.

The operational principle will be made clear if the diagram is studied along with this description. The plastic fuel pipe is connected to the petrol inlet 1, and from here fuel passes to the float chamber by way of the needle valve seating 5 and needle 8, the latter being

---

Fig. 4.15. Constructional details of Stromberg type C D carburetter.
1 Fuel inlet to float chamber. 2 Top cover (suction chamber) screws. 3 Throttle-stop idling screw. 4 Fast idling adjusting screw. 5 Needle valve seating. 6 Lever and cam for fast idle. 7 Float lever pivot. 8 Needle valve. 9 Lifting pin for air-valve piston. 10 Jet needle retaining screw. 11 Jet base O-ring. 12 Jet holder. 13 Jet screw for adjustment. 14 Air-valve damper piston. 15 Air-valve return spring. 16 Diaphragm. 17 Air-valve guide rod. 18 Air-valve piston. 19 Jet orifice. 20 Starter bar. 21 Fuel inlet to jet base. 22 Fuel inlet to jet bore. 23 Jet orifice bush. 24 Suction chamber. 25 Air passage to suction chamber. 26 Main air intake. 27 Throttle valve butterfly. 28 Bridge. 29 Jet needle.

THE INDUCTION SYSTEM AND CARBURETTER 65

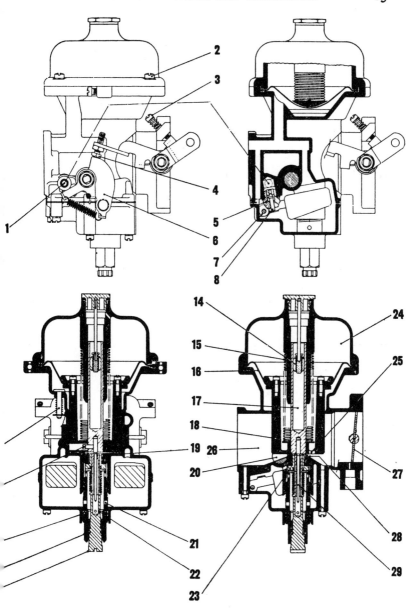

controlled by twin expanded rubber floats mounted on the arm 7. From the float chamber, the petrol passes to the jet orifice 19 via holes 21 and 22 in the jet assembly, the level being maintained at the same height as in the float chamber, in the usual manner.

Starting and idling

When the choke control on the dashboard is operated, it actuates the lever 6 at the side of the carburetter, this rotating the starter bar 20 to lift the piston-type air-valve 18. This valve has a metering needle 29 at its base, which passes down the jet orifice to control the effective bore of the jet. Thus when it is raised in the foregoing manner, the mixture is enriched for cold starting. Simultaneously with this, a cam on lever 6 opens the throttle beyond the normal idling position, as adjusted by means of the fast idle stop screw 4, this providing the faster idling speed required when warming up. As soon as the engine starts, the induction manifold vacuum will lift the air-valve 18, this weakening the mixture satisfactorily to prevent over-richness. The car can be driven away on this setting of the choke control, but the latter must be progressively released as the engine warms up, until the control is right out of action. The engine will idle in neutral at a speed determined by the throttle-stop screw 3. It will be seen that there is no separate idling circuit, the mixture being provided from the main jet and main air intake. The fuel quantity is determined by the actual height of the main jet in the carburetter body, and this can be adjusted by the adjusting screw 13 at its base. Turning the screw clockwise decreases the mixture strength, while anti-clockwise will enrich it. The throttle-stop screw 3 is adjusted to obtain a satisfactory idling speed. It is important to note that the throttle pedal must not be depressed when starting from cold.

Normal running

Under running conditions, the manifold vacuum is transferred by way of the passage 25 in the air-valve, to the suction chamber 24. This chamber has a diaphragm 16 at its base, which seals it from the main body. It will be evident that the pressure difference between the chamber 24 and the air-intake 26 (atmospheric) will cause the diaphragm, and with it the air-valve, to be pulled upwards towards the suction chamber. This action will not only increase the bore of the carburetter through-way, but will also draw the metering needle out of the jet, so allowing more fuel to pass. The lift of the air-valve is in proportion to the weight of air passing through the throttle valve 27, and as the jet needle is graded along its length to very close limits, the mixture ratio is maintained in line with the requirements of the engine, at any combination of rev/min and load.

In the suction chamber, it will be noted that a hydraulic damper piston is fitted inside the top guide rod 17. By preventing the sudden upward movement of the air-valve when the throttle is quickly opened, a richer mixture than normal is provided temporarily for assisting rapid acceleration. (This device takes the place of the accelerator pump on a fixed-choke carburetter.) The damper piston is shown at 14, and is suspended from the top screw of the suction chamber. It operates in oil contained in the hollow guide-rod, this being filled to within 1/4 in. of the rod end with oil of SAE 20 viscosity.

The air-valve is lifted against a helical compression-spring 15, which has to be specially graded. Apart from the float-needle valve, this spring and the jet needle are the only two components requiring selection.

Adjustments

For adjusting the idling, particularly after overhaul when the original setting may have been lost, the air-valve 18 is held down on to the bridge 28 in the throttle bore (the air cleaner being removed). The jet adjustment screw 13 is then screwed upwards with a coin, until the top of the jet can be felt to contact the underside of the air valve. From this position, turn the jet screw the other way, *i e* downwards, for three turns. This provides a good approximate position from which to start.

The engine is now run until it is at normal temperature, and the idling adjusted by screw 3 to a satisfactory figure; do not attempt to obtain an excessively slow speed. At the same time the jet screw 13 should be turned slightly to obtain smooth and regular firing, but only a little movement should be called for. When screwed upwards the engine may speed up initially, but further movement will cause it to stall. From this point, the screw must then be turned downwards, until firing is even at the desired slow-running speed. The air-valve 18 should now be lifted (via the air-intake) by means of a wire spoke or similar instrument, for a very short distance, say 1/32 in. or a little more. If the speed rises, the mixture may be rather too rich, but if the engine stalls it is too week; as a rule it is best to err very slightly on the rich side. When the idling speed is correct, it follows that the mixture range throughout the throttle movement will also be in order, *i e* with the correct needle and air-valve spring fitted.

Float level

The float level is correct, when, with the carburetter inverted, the highest point of the float is approximately 3/4 in. above the face of the main body, with the fuel needle valve on its seating. If the level needs to be reset, this is done by bending the tag that contacts the end of the needle 8. This tag must be at right angles to the needle when the latter is in the closed position. The level can be lowered by fitting an extra washer under the needle valve assembly, and this is a simple way of obtaining a small change in level without any bending of parts.

Constructional points

It will be apparent that movement of the air-valve and jet needle must be completely unobstructed. Such obstruction can be caused by the needle binding in the jet, but the construction allows for centralisation of these parts, as there is an annular clearance around the orifice bush 23. This allows the bush and jet to be positioned laterally, and then clamped in the most suitable position to ensure free needle movement in its bore. To check that the movement is free, the air-valve can be lifted by the pin 9, and its descent checked. If the movement is defective, and centering is required, the air-valve 18 must first be lifted, and the jet assembly 12 fully tightened. The jet is screwed up until its tip 19 is just above the bridge 28. The jet assembly 12 is now slackened off about half a turn, this releasing the orifice bush 23. Now allow the air-valve 18 to drop so that the needle re-enters the jet. This action will automatically centralise the jet. If the air-valve is reluctant to drop, it can be assisted by removing the top cap and damper from the suction chamber, and pushing the air-valve down with a copper rod from above.

The jet assembly 12 is now slowly tightened, with frequent checks to make sure that the needle is still quite free. This is done by lifting the air-valve about 1/4 in. and letting it drop. It should at all times abut firmly on the bridge, with no prior checks to movement. If at any time the air-valve itself shows signs of sticking this is probably due to dirt on its exterior. The valve assembly is removed by undoing the screws 2 and taking off the top cover, when the assembly, complete with its diaphragm can be lifted off the main body. The air-valve and the carburetter bore should be cleaned with paraffin, also the diaphragm. Petrol may be used, but this causes expansion of the diaphragm which must then be allowed to dry thoroughly before refitting, so that it will locate properly.

Obviously, if the jet needle is bent, free movement is impossible. The needle must therefore be handled carefully at all times. When fitting, its lower shoulder must line up with the lower face of the air-valve; the locking screw 10 must always be fully tightened. The diaphragm has a beading and locating tab moulded to both its inner and outer radii, this ensuring correct positioning. It is secured to the air-valve by a ring, held by screws with lockwashers. When assembling, the beading must be correctly located and the screws fully tightened. For the outer edge of the diaphragm a location channel is provided at the top of the main body of the carburetter, and it is important that the beaded edge of the diaphragm is properly engaged therein. When the suction chamber cover is fitted it must be positioned so that the screw holes line up with those in the main body, so that the diaphragm will not be displaced when the cover is assembled.

The air-valve rod and its guide should be kept clean, and whenever dismantled, a few drops of light oil should be applied to the working surfaces.

There is a rubber O-ring fitted between the float chamber spigot boss and the jet assembly. This must always be fitted correctly; if leakage occurs in those conditions the ring should be replaced.

Fixed-choke carburetters

The two carburetters already described operate by varying the choke or through-way area, the depression (vacuum) in the bore remaining constant. The alternative fixed-choke principle operates with an unchanged through-way area, the mixture quality being maintained by variation of pressure in various passageways. There is a wide variety of fixed-choke carburetters fitted as standard to popular cars, but the principle of operation, and the general requirements for satisfactory performance, cover all conventional variations. Two of the simpler types will now be dealt with.

Solex carburetter

This is shown in downdraft form on Fig. 4.16. The main components of the carburetter are static, and comprise a main jet which

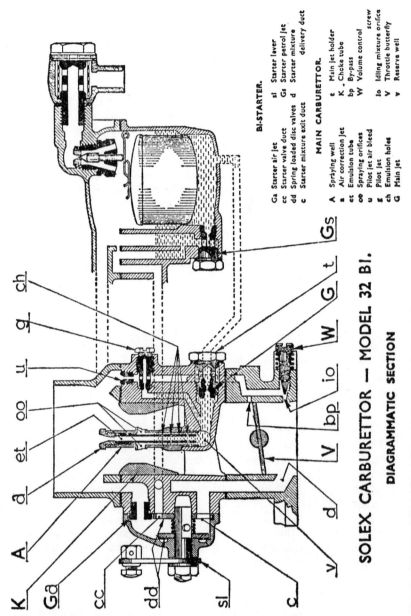

Fig. 4.16. Sectional view of Solex carburetter and 'bi-starter' device.

meters fuel from the float chamber into a horizontal passage leading from the main jet to a well, from which rises an emulsion tube. The latter is often mistaken for the main jet, but it will be evident that the jet is actually submerged all the time and operates unseen. At the top of the emulsion tube is an air-correction jet which also serves to lock the emulsion tube into position. It should be noted that the emulson tube faces 'backwards' from the throttle butterfly so that as the air stream flows past, it hits the air-correction jet orifice first. The operation is that the air stream past the emulsion tube enters (in part) the air-correction jet. At the same time, fuel from the main jet is passing into the well at the bottom of the emulsion tube. The air and fuel-streams meet, forming an emulsion which then passes to four spraying orifices located about half-way down the emulsion tube, and almost centrally in the choke-tube, or venturi, which encircles the emulsion tube. From the spraying orifices the emulsion meets the main air stream, and passes via the throttle valve to the engine.

The idling mixture is supplied from the well at the base of the emulsion tube, by way of a channel, through a pilot jet and thence into an idling orifice which is controlled by a tapered adjustable screw-valve. This orifice is actually on the engine side of the throttle valve and is thus unaffected by it. However, there is also a by-pass duct connecting the idling orifice to a point just on the other side of the throttle, when the latter is in the almost-closed position.

Assuming that the engine is idling, the throttle being nearly shut, the by-pass duct acts as an air-bleed on the idling fuel supply, thus preventing over-richness. As soon as the throttle is opened, the throttle butterfly passes the orifice opening, which is in effect transferred to the atmospheric side of the carburetter, so that both the idling orifice and the by-pass duct now act as fuel delivery inlets. In so doing, they proportionately enrich the fuel output at the transfer position between pilot and main supplies, and prevent a period of throttle travel with the mixture unduly weak, or a 'flat spot' as it is popularly called.

Starting mixture

Operation of the 'choke' control on the Solex causes an auxiliary device to come into action. The usual version of this, known as the 'bi-starter', gives two positions of the control; one, very rich, for cold starting, the other, an intermediate, for allowing the vehicle to be driven away without the possibility of over-richening and choking with neat fuel. The bi-starter comprises a chamber containing a disc valve, the latter operated by the choke control. The chamber is fed with fuel from its own submerged jet, and with air via its own air jet. The resulting mixture from the chamber (which is in effect an auxiliary carburetter for starting only) is fed to the carburetter body at a point on the engine side of the throttle valve, by way of a large duct. The duct is, of course, put into communication by means of the disc valve, and the alternative positions of the latter decide whether the fully-rich or intermediate-rich mixture is supplied.

It will be evident from the above brief description, that the correct functioning of this type of carburetter depends on the selection of several metered orifices, or jets, by size. The starting device has one fuel and one air jet. The main carburetter has its main jet, pilot jet, and air-correction jet.

Adjustment of the Solex carburetter

As far as the starting device is concerned, the air and fuel jets are chosen by experiment to suit the engine when delivered, and it is most unlikely that any change will be called for. In fact, the air jet is selected on a basis of engine capacity, and though a change may be required under some conditions of climate or altitude, haphazard selection is unwise, and the maker's advice should be sought in such cases.

As regards the fuel jet in the starter, this can be checked simply enough. Starting from cold should be practically instantaneous, and the engine should keep on running after the initial start. If it does not, and is otherwise in good order, a size larger fuel jet should be tried.

If after starting, the exhaust emits black fumes and the engine tends to eight-stroke, or 'hunt', this is a sure sign of too rich a mixture. Confirmation is given if these signs persist even when the control is returned to the intermediate position. In such cases a smaller jet should be tried. It will be realised that the above adjustments to the starting part of the carburetter are not over-critical, and it is when the 'real' carburetter is tackled that great care is necessary.

The idling adjustment is controlled by the normal type of throttle-stop screw on the throttle butterfly actuation. The throttle is opened by screwing 'in', thus speeding up the engine. The idling mixture strength is controlled by the tapered screw aforementioned, turning anti-clockwise for enrichment and clockwise for weakening the mixture. The fuel output for idling is, of course, supplied by the pilot jet, which thus fixes the limit of richness of this portion of the mixture.

The technique of adjusting these items to obtain correct idling will be described later. Before doing this the remaining adjustable items on the carburetter will be described.

Firstly, the choke tube, or venturi. This is always selected to suit the engine to which the carburetter is fitted and should never need altering unless some very drastic modifications to the power unit have taken place.

The main jet, also, is normally selected to match the choke tube size and is tuned and corrected by the use of a suitable air correction jet which, as already described, is located at the top of the emulsion tube.

Selection of part sizes

Whenever an engine lacks performance in some respect, there may be considerable temptation on the part of certain owners to experiment with the carburetter jet sizes and so on. It should be stated clearly that, as far as carburetters of reputable make, of the fixed-choke type are concerned, these, when fitted as standard to any particular engine, have their jet and choke tube sizes selected by road and bench test, and no advantage can be expected from altering the standard selection.

There are, however, cases where the engine's characteristics may have been modified, or where fixed-choke carburetters are being substituted for different types. In these cases it is useful to know the significance of various changes of components, and the following description of the functioning of the parts will be of assistance.

If the engine is assumed to be at rest and the fuel supply system in order, the spraying assembly will be filled with petrol to a point near the spraying orifices. When the engine is started and air flows in the choke tube and past the spraying assembly, petrol is drawn up due to the increasing vacuum in the narrow diameter of choke tube. It will be obvious that if this was all that happened the mixture would become progressively richer as the air speed rose in the choke tube. This degree of richness must therefore be adjusted to be right at all values of air speed in the choke tube, and this task is done by the emulsion tube, via the air correction jet which is screwed into its top, facing the incoming air. This air correction jet, it will be realised, regulates the amount of air which is 'tapped off' the main air stream to pass into the emulsion tube and to emerge from the spraying orifices along with the petrol.

This correction air, having mixed with the petrol in the emulsion tube, reduces the mixture strength in a progression which is directly opposed to the increasing mixture strength which would be obtained if there were no correction, *i e*, if a simple jet and choke tube combination only were involved. It does this in two ways: firstly, by progressively relieving the air vacuum and, secondly, by the mechanical 'obstruction' which it causes to the fuel flow. This latter is, of course, the result of the correction air and the fuel flowing in opposite directions towards each other, but this feature of the mixture strength correction system has other virtues. The main one, in fact, is considered to be that the mixture strength can be altered at one part of the operation curve without affecting other parts. For example, if it is required to enrich the mixture on the lower part of the curve (low speed mixture) the main jet size is increased without altering the air correction jet size. Conversely, the mixture strength at the top of the curve (high speed) can be increased by decreasing the size of the air correction jet without altering the main fuel jet.

Examples of jet selection

To make the matter clear, we can take one or two practical examples. Considering an engine which is using a combination consisting of size 25 choke tube, main jet 120, and air correction jet 240, we suspect that the main jet is too large, since a size 115 gives good acceleration and flexibility. It is, however, found that with a 115 jet fitted there is a falling off of power at the higher end of the power range, indicative, of course, of too weak a mixture at this point.

To correct this weakness high up we need to reduce the size of the air correction jet which at the moment is 240. We try size 220 or 200 and find the results equal to those obtained with the original combination as far as power output throughout the power range is concerned. Since, however, the main jet size has been reduced from 120 to 115, it will obviously be passing less fuel, so that the same performance will be available with greater economy.

To take an example of the opposite situation we might have an engine using a combination of 25 choke tube, 115 main jet and 240 correction jet. The performance is good except that there is a definite flat spot and acceleration is poor.

Both the defects point to an excessively weak mixture at the bottom end. We therefore increase the main jet size to 120. This corrects the flat spot and benefits the acceleration, but the consumption is excessive, particularly at high speeds. It is thus evident that the mixture at the top end is too rich. A larger air correction jet, say 260 or 280, must be fitted, which suitably weakens the top end mixture. The fuel consumption will then be restored to normal.

It will be seen from the foregoing that the assembly of main jet and emulsion tube is capable of alteration to cater for all required conditions of engine operation. As already remarked, however, experimentation should not be necessary except for special purposes.

The method of obtaining mixture for idling has already been described. To increase or decrease the idling speed it is only necessary to adjust the throttle-stop screw until the desired speed is obtained. However, this does not always result in the engine firing evenly, due to the mixture being incorrect. It is usually necessary, therefore, to experiment with the mixture adjustment screw, in conjunction with the throttle-stop screw, turning it anti-clockwise for enrichment and clockwise to weaken the mixture.

Over-rich idling mixture is shown, as usual, by lumpy running and black exhaust fumes. A weak mixture shows itself by a tendency for the engine to stall, and misfiring. There is no difficulty about obtaining perfectly even idling so long as the mistake is not made of trying to get the engine to 'tick-over' at a ridiculously slow speed. For the modern o h v engine, 600 to 800 rev/min is by no means too fast. Any idea that fuel will be saved by lowering the speed from this figure is quite erroneous.

General maintenance

Although the preceding description of the various possible adjustments may give an impression of complication, it should be appreciated that for all normal purposes the carburetter as supplied by the makers for the engine, will be fitted up with correct jets and so on. It is only when special conditions have to be met, that a change may be necessary, and in such cases it is essential to know just what is involved.

The need for economy in fuel leads many drivers to consider the possibility of gaining something by fitting smaller jets. It is, of course, just possible that a little more economy might result from such an action, but the deterioration in performance is most undesirable. Obviously, for any particular bore, or choke size, of carburetter, the range of jet sizes must be limited, and within these limits the best performance in regard to power output will be obtained. If extreme economy, to mean anything at all, is required, even at the expense of power, the answer is to fit a smaller carburetter, *i e*, one having a smaller choke and through-way.

Flooding of the carburetter while in operation is usually caused by dirt on the float-needle seating, but this should not occur providing the filters are in good condition. The Solex needle and seating complete is renewable in much the same manner as described for the S U instrument. It is, however, sometimes possible to correct a slight leakage between needle and seating by removing the assembly from the carburetter, inverting it, and lightly tapping the needle into the seating a few times, rotating the needle between the taps. The effect of this is to form a new seating.

Too high a fuel pressure can, of course, lead to fuel being forced past the needle. It is not often that mechanical or electric fuel pumps increase their delivery pressure in use; the contrary generally happens. The delivery pressure can, however, be readily checked with a gauge applied to the delivery line, and should not exceed just over 2 lb per sq. in.

Air leaks at the throttle spindle may occur after a very long mileage, and replacement of the carburetter by an 'exchange' instrument is the most satisfactory remedy. As in the case of the S U, a little oil applied to the spindle will make for freedom from this complaint.

As the jets constitute fixed orifices, it is possible for these to become choked, if the filters are in such poor condition that foreign matter can pass through. A choked or partially choked jet is easily diagnosed by its effect on performance, which may vary from merely erratic running to a complete shut-down. The quickest remedy in the long run is removal of the offending jet, and a proper cleaning. Short cuts such as blowing through the passages with a tyre pump are very seldom successful, as there will probably be plenty more dirt about the place. Needless to say, jets should only be removed with a jet-key designed to fit them.

Unless absolutely unavoidable, jets should never be cleaned by poking wire through them. Bearing in mind that their orifices are only a few thousandths of an inch in diameter it is obvious that forcible cleaning in this manner could actually ream out the hole. The correct way is to use a tyre pump or similar method of obtaining a blast of air.

The Zenith carburetter

Another widely-used type of fixed-choke carburetter is the Zenith. There are several variations made, but the basic working principle is the same for all. If the sectional drawing (Fig. 4.17) is studied in

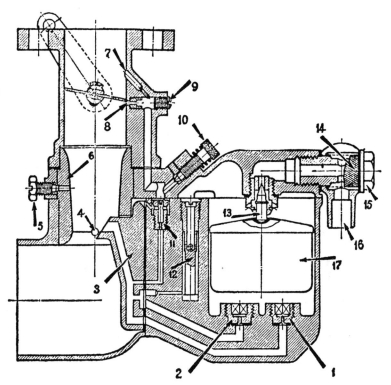

Fig. 4.17. Sectional view of Zenith type V carburetter.
1 Main jet. 2 Compensating jet. 3 Emulsion block. 4 Distributor bar. 5 Choke retaining screw. 6 Choke tube. 7 Slow-running outlet. 8 Progression jet. 9 Progression jet cover or plug. 10 Slow-running airscrew. 11 Slow-running jet. 12 Capacity tube. 13 Needle. 14 Filter. 15 Filter union. 16 Petrol union. 17 Float.

conjunction with the following description, the functioning can be readily followed.

The first thing to notice is that two jets are located in the base of the float chamber or 'bowl', as it is called. Assuming that the chamber is filled (via the usual needle-valve mechanism) fuel will pass through the main jet and the compensating jet, these being the two mentioned. From these jets, the flow is along passages to a common channel in the emulsion block, which projects into the main intake.

When starting from cold, rich mixture is provided by a strangler flap in the main air intake. On some types the flap is interconnected with the throttle so that when the strangler is operated the throttle is slightly opened, thus relieving the driver of the necessity for gauging the correct amount of throttle to be given for starting. Another type of strangler is semi-automatic, so that if it is incorrectly kept in operation after the engine has increased speed the intake vacuum partially opens the strangler flap, and prevents an undesirably rich mixture from being inhaled.

There are other starting devices, including a fully-automatic type that supplies an extra rich mixture through a passage on the 'engine' side of the throttle without the use of a strangler flap. This needs no skill in operation, but no throttle movement is necessary, or permissible, when starting, whereas with the flap type of mixture enrichment, a little extra throttle is required.

Working principle

Assuming that the engine is at working temperature and idling speed, the throttle will be closed and the engine vacuum concentrated on the slow-running outlet, connected to the slow-running jet. Fuel will be drawn from the well beneath the jet, being metered in the jet and then meeting air entering at the base of the slow running air screw. The amount of fuel issuing from the jet is thus controlled by this screw. At the edge of the throttle, there is a further outlet from the progression jet, this breaking into the slow-running passage. When the throttle is opened up from the idling position, this outlet gives an additional mixture to ensure progressive getaway from slow running.

When the throttle is still further opened, the vacuum will be concentrated on the 'beak' of the emulsion block which, it will be noticed, projects into the narowest part of the choke tube. The result will be, firstly, that fuel will be drawn from the main channel in the emulsion block, also from the channel below the slow-running jet, and from the well of the capacity tube. Thus, the source of fuel supply must eventually be metered via the main and compensating jets. The capacity tube is open to atmosphere at the top, so that when the fuel in its well has been used, any more fuel coming from the compensating jet along the passage beneath is mixed with air. By this means, mixture correction is maintained. As the vacuum increases, the compensating jet supplies a weaker mixture, while the main jet delivers more fuel, which meets the emulsified fuel from the compensating jet in the common channel. This tends to break up the main jet stream, and thus the supply issuing from the emulsion block beak into the choke tube is completely atomised.

A distributor bar will be noticed across the choke tube, at right-angles to the beak. This is fitted in order to ensure that the mixture is distributed across the choke tube in all directions, the supply issuing from the beak striking the bar.

Zenith adjustment

The Zenith carburetter is very simple to dismantle, given reasonable care. The bowl, containing the float and jets, is removable without disturbing the fuel pipe connection, merely by undoing its two fixing screws. Hold the hand below the bowl during this operation, and catch it as it drops.

After removing the float, the jets are in view. One of the two bowl-fixing screws is specially shaped at its end to act as a jet-key. This end should be inserted into the jet, and the screw head turned with a spanner. The slow-running jet, which will appear as being on the edge of the bowl, can be removed with a screwdriver, as also can the screw holding the capacity tube. The latter will drop out if the bowl is inverted. (If there does not appear to be any capacity tube fitted, don't worry; your type of instrument has a cored passage instead.)

The emulsion block is held to the side of the bowl by three or five screws. There is a correct method of removal; the bottom screw is just eased, and then the remaining screws are taken right out. The bottom screw is then turned anti-clockwise, when it will come away complete with the emulsion block. Take care not to damage the gasket between the block and the bowl, and never use jointing compound on the gasket. When replacing the block, locate the bottom screw first, then tighten the others evenly.

The starting and progression jets can be removed with a screwdriver; the plug over the progression jet which is shown in the diagram is frequently made in one with the jet, forming a complete unit. When cleaning jets, nothing should be passed through that is liable to enlarge the orifices; an air-blast will usually be found sufficient to remove any obstruction.

Slow running

The usual throttle stop screw is provided to adjust the throttle position for idling, a clockwise movement increasing the speed. The richness of the slow-running mixture is adjusted by the slow-running air screw. Erratic idling, and stalling on deceleration, points to a blocked slow-running jet. 'Lumpy' idling indicates too rich a mixture, and this can be weakened by turning the air-screw anti-clockwise. The right position for this screw is about one full turn out from the fully-home position, except in some cases where the screw has an elongated taper end; for these, the position is about three turns. If correct idling cannot be obtained with the air-screw very near this position, a diffierent size of slow-running jet is indicated. The throttle-stop must, of course, be adjusted in conjunction with the air-screw to obtain the right idling speed.

It will be appreciated that the foregoing adjustment affects also the acceleration, and if maladjusted, will cause a dead spot when opening up from idling; for this reason, an adjustment a shade on the rich side is advantageous. The compensating jet size governs acceleration through the speed range; with too small a compensating jet, the engine will pause appreciably before responding, and some spitting back may occur. If the jet is too large the acceleration will be laboured. If acceleration still fails to satisfy after adjusting the slow-running, and perhaps trying different compensating jets, a different size capacity tube can be tried (if fitted), though this is unlikely to be called for.

When pulling hard at low speeds, mixture strength is again determined by the compensating jet, and if power is lacking on hills, the jet should be changed. The main jet looks after the high-speed output, and if, with all else in order, it is felt that the carburetter is holding back on power, it is worth trying a larger main jet.

Faults and their causes

A poor get-away, and possibly popping-back, points to too small a compensating jet. Lack of power at all speeds, again accompanied by popping-back at intervals, indicates too small a main jet. If explosions occur in the silencer when running downhill, the slow-running mixture is too weak.

Heavy fuel consumption may be caused by having the jets too large, but, of course, if the performance is right, the fuel metering must also be taken as correct. Smaller jets can be used for economy, but power will suffer. Excessive usage of fuel is more likely to be caused by the strangler flap not returning to the open position after use, or to a defective automatic starting control.

A filter gauze is fitted at the point where the fuel pipe enters the needle valve housing in the carburetter casting. Removal of the pipe union will give access to the gauze, which should be washed occasionally in petrol. A certain amount of sediment will collect in the bowl, and although the jets are raised clear of the bottom, too much debris should not be allowed to accumulate, as otherwise it will tend to be drawn through, causing partial blockages which can be irritating. It is worth while inspecting the bowl every 10,000 miles or so, and if necessary, washing out the sediment.

Multiple carburetters

Many cars have more than one carburetter and in such cases it is necessary to ensure that their operation is properly synchronised.

With fixed-choke carburetters there is little difficulty, so long as the carburetters are identical in all respects. It is only necessary to ensure that the control linkage is working in such a way that the two throttles move absolutely in step. An exception to the foregoing is sometimes found in the case of six-cylinder engines having three carburetters. Due to the regular induction impulses from the two cylinders on the middle carburetter (in contrast to the irregular action of the outer pairs), the former may call for a different setting aimed at a slightly weaker mixture.

In the case of twin S U and Stromberg carburetters, because of the variable-jet feature, correct synchronisation is rather more complicated, although the procedure is quite straightforward. The tuning sequence already described is simply adapted to cover the two carburetters, and is as follows.

Examine the action of the two throttles and their interconnecting coupling, and make sure that both throttles close completely when their slow-running adjustment screws are slackened right off. Slacken the clamp bolt of the coupling and check that the throttle action is quite free. Then screw the slow-running screws inwards for about $1\frac{1}{2}$ turns, and see that the two throttles open slightly to the same amount and are held by their stop-screws; then tighten the coupling.

On S U carburetters, the two jet-adjusting hexagons must now be screwed upwards to their topmost position, but not tightly as they must be easily movable. Start the engine; it will be necessary to pull down the jets by the mixture control to do this and a degree of throttle opening will also be necessary. With the engine started, set the throttles by the slow-running screws to a steady 1,000 rev/min. Gradually move the jet control to the topmost position of the jets as the engine warms up. Next, with a cycle spoke, very carefully lift the air-piston on the first carburettor about $\frac{1}{16}$ in., taking care not to obstruct the air intake while doing so. It will be found by this action that the two cylinders served by the carburetter will misfire and the engine will tend to stall. Do likewise with the other carburetter, and the same result will be noted.

Now go back to the first carburetter again, and screw the hexagon down, that is, clockwise, half a turn. Try the action with the spoke again. This time it may be found that the engine does not misfire, which is what we are aiming at. Apply the same treatment to the second carburetter, moving the jet hexagon the same amount on each instrument. Test between each movement, until a position is arrived at where no effect is shown in the running of the engine when the air-piston is lifted (taking care not to lift it more than the fraction mentioned).

It will probably be found that the engine speed is tending to drop, and so the idling screws should be reset to give 1,000 rev/min as before. Next, move each hexagon one 'flat' or $\frac{1}{6}$ of a turn clockwise. We should now find that lifting the air-slide causes the engine speed to increase slightly, showing that the mixture is very slightly on the rich side when idling; this is the object, and it may be found that a very slight difference in the positions of the two hexagons may be needed to obtain the same speed increase on each carburetter. The difference, however, should not amount to more than one 'flat'.

In the case of Stromberg carburetters, synchronisation procedure is similar to the foregoing. The two throttles are first checked for closure in unison, and each of the stop screws is adjusted until its tip is just touching on the throttle spindle stop lever.

From this point, turn each screw one complete turn, ensuring that each one is taking the load of its individual throttle, *i e* that there is no load imposed on the interconnecting spindle in consequence of the two screws being adjusted differently. Each carburetter should then have its jet screw adjusted as already detailed. With the engine at running temperature the procedure for setting the idling is as previously described. Each carburetter is treated individually except that the throttle-stop screws must be adjusted in unison for the reason stated above. When applying the piston-lift test with the spoke, the effect will be noted on the two cylinders served by that particular carburetter, and the aim must be to obtain an identical effect on each pair of cylinders.

Aural Check

As a final check for both types of carburetter, listen to the hiss from each intake, which should be equal in intensity; a rubber tube can be used for this purpose. Also listen to the exhaust note. A splashy beat indicates weak mixture and a heavy thumpy note shows over-richness, the latter being accompanied by black smoke. However if the tuning sequence has been properly carried out, all should be well and, at 700–1,000 rev/min (depending on size of engine), the idling should be quite regular. In very hot weather it may be possible to weaken the mixture to a small degree, but when doing this always be careful to move both jet adjustments in turn, and very little at a time. If one is moved drastically without the other, the original settings will be hopelessly lost.

On triple-carburetter installations for six-cylinder engines, the same procedure is followed, the three being tuned in numerical sequence. As already stated, it may be found that the central instrument will take a slightly weaker mixture than the two outers. Thus there need be no anxiety if the jet position (or size on fixed choke types) is found to differ from that of the other two.

On some twin-carburetter installations the interconnecting shaft between the two throttles has a type of coupling arranged to give a 'lag' on one throttle, so that it opens slightly behind the other one. The object is to improve smoothness of acceleration from low revolutions, by concentrating the induction depression on one choke only. When adjusting the interconnection, the lag must be maintained but otherwise the procedure for synchronisation is unaltered.

Reliability

We mentioned earlier that it has been fashionable from the earliest days to blame the carburetter for many of the ills to which the engine is heir during its life. Why this should be so is rather hard to grasp, since readily obtainable figures show that as far as breakdowns are concerned, ignition faults account for about 80 per cent of these. This is no fault of the ignition system, but merely shows that the sparks department will not stand the same amount of neglect without protesting. Carburetters, being relatively much more simple, are, in fact, extremely reliable. Where engine 'trouble' includes high fuel consumption, poor acceleration, misfiring, and so on, singly or in combination, the carburetter should be the last to be blamed, always providing it is receiving clean fuel.

It should be borne in mind that an engine in regular use must progressively deteriorate in performance. Whether it does this quickly or slowly depends very largely on the general standard of maintenance. If it is neglected, it may develop symptoms which could be diagnosed as indicating faulty valves, ignition, or carburetter, with almost equal reason. The last-named is, however, the least likely.

CHAPTER 5

Fuel Supply Apparatus

The S U electric pump—Delivery to order—Contact-breaker—Dismantling the pump—The valves—Checking the valves—The diaphragm—Stretching the diaphragm—Testing the pump—The contact-breaker assembly—Adjusting the points—The P D electric pump—Filter cleaning—Faulty action—The A C mechanical pump—Overhauling—Diaphragm refitting—Fuel Connections

The next item to receive consideration is the fuel supply apparatus. It has already been mentioned that blaming the carburetter is rather fashionable, wherever in the engine the fault actually lies. This is particularly so in the case of the petrol supply system, as faulty operation due to an intermittent head of fuel often gives rise to symptoms which suggest something wrong with the 'gasworks'.

The S U electric pump

The S U electric fuel pump is fitted to a large number of cars, and will be dealt with first. Its action is independent of the engine, by energy supplied from the battery. In consequence it is possible to pump up a head of fuel in the carburetter, to compensate for evaporation, before the engine is rotated, thus saving some wear and tear on the starter motor, and drain on the battery. The pump is usually wired via the ignition switch, its characteristic clicking noise when the engine is switched on indicating that the carburetter is being filled. This noise is also useful for tracing faults, as will be shown later.

The description given applies to the S P and A U F type pumps which are widely used. Other types operate on the same principle but may have the accessories, valves, etc, arranged in a slightly different manner. This however does not affect the method of working, nor the general requirements for overhaul.

A sectional view through the pump is shown on Fig. 5.1 while the components are itemised on Fig. 5.2. The main assemblies comprise the pumping chamber, the magnet body, and the contact-breaker. The pumping chamber is provided with tapped holes for receiving screws which attach it to its mounting. It is possible to do such jobs as cleaning the pump filter without removing the pump, so long as the latter is mounted in a reasonably accessible position. Filter cleaning is however only required at quite long intervals, say up to 10,000 miles, so long as clean fuel is used at all times. At about double this mileage, or a little less, it is a good idea to give the other pump items a check over, with the pump removed completely from the car. To do this it is only necessary to detach the inlet and delivery pipes and the electrical supply and earth leads. Removal of the fixing bolts will then allow the pump to be withdrawn. In cases where the pump is mounted in the boot of the car, in proximity to the tank, it may be located behind a trim panel, and this must be removed first. If the pipe unions are of a type which screw into junctions that are themselves threaded into the pump body, the junctions should be held with a second spanner while undoing the union nut, as otherwise the action may twist the pipe and lead to fracture.

Assuming that the pump has been removed to the bench, an examination will show the working principle.

The suction and delivery unions may take various forms, but in all cases a filter gauze is incorporated in the former, and is easily removed after detaching the union. The suction and delivery valves are located at the back of the pumping chamber and are exposed when the diaphragm has been removed. Before describing the dismantling operation the action of the pump will be dealt with.

The circular side of the pumping chamber remote from the fixing holes is closed by a flexible diaphragm held by a ring of slotted screws. On the outer side of this diaphragm is the magnet body housing, which encloses a wound coil, or solenoid. At the outer end of the magnet body is fitted the contact-breaker, covered by a moulded cap which is held on by the electrical terminal.

At the end of the magnet coil is an armature of iron, so that if the coil is energised it will pull the armature axially. The armature is attached to a non-magnetic push-rod which is in turn attached to the centre of the pump diaphragm at one end, and at the other end engages the contact-breaker mechanism. Thus, movement of the arm-

---

Fig. 5.1. Sectional view of S U electric fuel pump.

1 Body. 2 Coil housing. 3 Contact-breaker assembly. 4 Contact-breaker pedestal. 5 Toggle springs. 6 Outer rocker. 7 Coil. 8 Magnet core. 9 Return spring. 10 Earthing screw. 11 Diaphragm. 12 Delivery nozzle. 13 Delivery valve. 14 Clamp plate. 15 Clamp screws. 16 Filter. 17 Suction nozzle. 18 Suction valve. 19 Pumping chamber. 20 Rollers. 21 Vent to atmosphere. 22 Armature. 23 Armature spindle. 24 Trunnion. 25 Rocker spindle. 26 Inner rocker. 27 Earth connection. 28 Terminal stud. 29 Spring blade. 30 Contact points. 31 End cover.

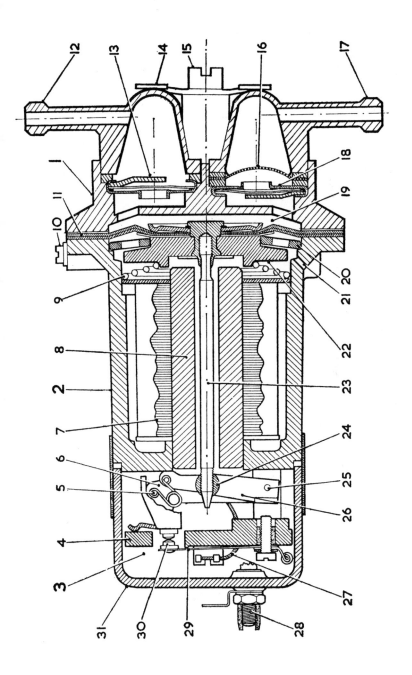

ature in an axial direction will actuate the diaphragm and the contact-breaker.

When the diaphragm is pulled outwards from the pumping chamber, it causes a vacuum inside the latter, performing the suction stroke. This is done electrically by energisation of the solenoid. As soon as the suction stroke is completed, the contact-breaker is operated by the aforementioned push-rod, and the current is cut off from the winding. The delivery stroke entails the return of the diaphragm in the reverse direction, towards the pumping chamber, so as to cause pressure inside the latter. This return stroke is achieved by a large coil-spring fitted between the diaphragm and the end of the magnet coil.

Delivery to order

It will be evident from the above that, if the carburetter float chamber is full, and no more fuel can enter, the pump will remain in the 'end of suction' position, with the current cut off, and the return spring fully compressed, exerting pressure on the fuel in the pumping chamber. As soon as fuel is required, the spring pressure will deliver it, and when the pumping chamber has delivered its full volume, the diaphragm push-rod, having moved its full stroke, will actuate the contact-breaker, making the circuit, energising the solenoid and once again performing the suction stroke to fill the pumping chamber ready for the next 'call'.

The pressure required to lift the petrol to the carburetter is only about 2 lb per sq. in., and any higher pressure would force the fuel past the needle valve and lead to flooding. The spring in the pump is thus carefully graded to build up just enough pressure and no more.

The efficiency of the pump as such, having regard to the above

---

Fig. 5.2. Components of S U electric fuel pump.
1 Pump assembled. 2 Body. 3 Filter. 4 Valve. 5 Banjo. 6 Sealing washer. 7 Clamp plate. 8 Clamp screw. 9 Coil housing. 10, 11 Terminal tags. 12 Earthing screw. 13 Spring washer. 14 Housing to body screw. 15 Diaphragm assembly. 16 Joint washer. 17 Return spring. 18 Roller. 19 Rocker and blade assembly. 20 Blade. 21 Terminal tag. 22 Blade screw. 23 Dished washer. 24 Contact-breaker spindle. 25 Pedestal. 26 Pedestal screw. 27 Spring washer. 28 Terminal stud. 29 Spring washer. 30 Lead washer. 31 Recessed nut. 32 Plain washer. 33 End cover. 34 Cover nut. 35 Shakeproof washer. 36 Connector. 37 Insulating sleeve. 38 Vent valve. 39 Valve ball. 40 Vent cap. 41 Sealing ring.

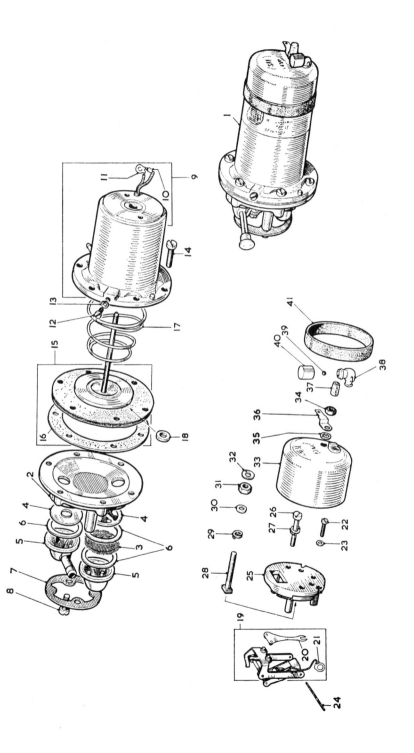

points, depends very largely on the valves. These are of a flat disc type mounted close together and positioned by a clamp plate, the precise details being given on the diagrams. On the suction stroke the diaphragm is pulled towards the magnet core, lifting the inlet valve off its seat and filling the pumping chamber with fuel via the inlet union; meantime the delivery valve is held firmly against its seat. On the delivery stroke the pressure set up by the returning diaphragm presses the suction valve firmly against its seat, and lifts the delivery valve, allowing fuel to pass to the delivery union.

The pump diaphragm is obviously not designed to carry the weight of the armature assembly at that end of the pump, and in order to carry the weight and at the same time allow the assembly to reciprocate freely, a number of spherically-edged brass rollers are arranged round the armature and located in a machined recess in the outer body.

Contact-breaker

The contact-breaker is exposed by removing the bakelite cover at the outer end of the pump. This is done by first unscrewing the terminal nut, and the insulated sleeve and connector. To check the contact and rocker action on pumps having the assembly as on Fig. 5.3 the outer rocker should be pressed to the coil housing. In this situation the contact blade should just rest on the ledge projecting from the main face of the pedestal. If it does not, the attachment screw of the blade must be slackened, the blade swung clear, and then bent downwards slightly so that when again in position it lightly contacts the ledge. If bent too much, the rocker travel will be restricted.

To check the rocker finger setting, indicated by gap 'A' on Fig. 5.3, hold the contact blade lightly against the pedestal ledge, taking care not to press on the blade tip, and check the gap with a feeler gauge. The gap can be reset if necessary by bending the blade.

On assemblies arranged as on Fig. 5.4 there are two gaps to check. That marked 'A' is between the blade and the pedestal, and can be adjusted if necessary by bending the stop-finger which locates under the pedestal. This stop is also used to adjust gap 'B', between the rocker finger and the housing.

Contact points should always be adjusted so that the points on the blade are just a little above those on the rocker when closed, so that when make-and-break takes place, one pair wipes symmetrically over the other pair. This adjustment is made possible by the latitude on the attachment screw (Fig. 5.5). After adjusting, make sure this screw is properly tightened.

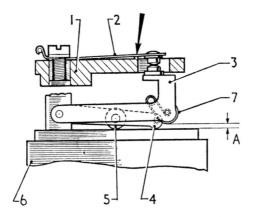

Fig. 5.3. Details of one type of contact-breaker assembly.
1 Pedestal. 2 Contact blade. 3 Outer rocker. 4 Inner rocker. 5 Trunnion. 6 Coil housing.
Gap A = 0·30 in.

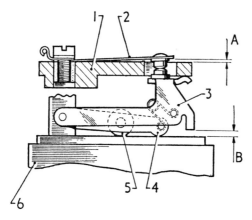

Fig. 5.4. Details of second type of contact-breaker assembly.
1 Pedestal. 2 Contact blade. 3 Outer rocker. 4 Inner rocker. 5 Trunnion. 6 Coil housing.
Gap A = 0·035 in., Gap B = 0·070 in.

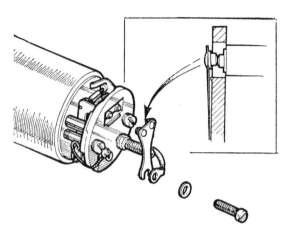

Fig. 5.5. Method of lining-up contact-breaker points.

Dismantling the pump

To inspect the valves and diaphragm, the pump must be completely dismantled; this is by no means difficult if ordinary care is used.

Six screws will be noted at the flange, which is the junction of the pumping chamber and the magnet outer case. These should be slackened off until they are all nearly unscrewed. A knife blade should next be very carefully run round the edge of the diaphragm to separate it from the pumping chamber body. When it is ascertained that the diaphragm is well and truly unstuck, the screws can be fully removed and the pumping chamber pulled away, leaving the diaphragm, complete with its rollers, still in position against the flange of the outer magnet case.

The pumping chamber body can be cleaned thoroughly in petrol, in ordinary circumstances. Occasionally, however, particularly when doubtful fuel has been used, it may be necessary to adopt stronger measures. A complete cleansing can be effected by using a 20 per cent caustic soda solution, followed by a dip in strong nitric acid and finally a good wash in boiling water. Take care when using these items, as they are not good for the hands. The method will remove the most obstinate and gummy deposits, and can be applied to all parts which come in contact with fuel, except, of course, the washers and the diaphragm itself.

The valves

Ready access to the valve assemblies is obtained in both types of pump. On the S P type, the suction union is first unscrewed, and the filter withdrawn. The two valves can then be taken out after undoing the Phillips screw. On the A U F pump there is a clamp plate, secured by 2 BA screws, which holds the suction and delivery stubs. If these are removed the valves can be taken out.

The valves should give no trouble unless unfiltered petrol has been allowed to enter, in which case a thorough cleaning should remedy matters. The filter should be washed in petrol periodically, but in normal conditions it should not require attention very often.

To reassemble the valves on the S P pump, the delivery valve is first fitted into its recess with the spring downwards and fitting evenly on the seating in the body. The clamp plate can then be refitted, making sure that the suction valve is central on the seating, and the Phillips screw tightened. When replacing the filter, insert it into the recess in the inlet union before screwing into the body; this reduces liability of the filter tip being forced into the valve recess and fouling the valve.

On the A U F pump, the delivery valve assembly is inserted with the tongue side uppermost into the recess marked 'outlet', and the delivery stub placed on top along with the washer. The suction valve is fitted similarly with the tongue side uppermost. Ensure that both the assemblies and stubs are firmly home, position the stubs to point in the required direction for lining up with the fuel piping, and refit the clamp plate, tightening the two screws firmly.

Checking the valves

After reassembly the action of the valves can be checked. To check the suction valve, attach a length of rubber tubing to the inlet union. It should be possible to blow freely down the tube, the air passing through the suction valve and emerging into the pumping chamber, which faces the diaphragm when the latter is assembled. No air should, of course, emerge from the outlet union, and it should not be possible to suck down the tube, as this action should pull the suction valve firmly down on to its seat. The delivery valve is checked in a similar manner, by attaching the rubber tube to the outlet union. In this case it should be possible to suck air through the tube (the air entering via the holes adjacent to the diaphragm aforementioned), but not to blow, as the latter action will close the delivery valve against its seat in the valve cage.

If the pump has been dismantled because of diaphragm failure as detailed below, the foregoing test on the valves can of course be done before taking them down, and if they 'pass', they should be left alone.

The diaphragm

The flexible diaphragm is usually the first part to give trouble, on a pump which is normally well maintained. This is no reflection on the item but merely indicates that it is an 'expendable' part. Its life is, however, about 30,000 miles, and impending trouble is shown long before the pump stops operations for good. Deterioration of the diaphragm shows up in the form of minute leaks, which not only cause the pump to operate unduly quickly without effectively pumping (due to passing air through the leaks) but eventually allow fuel to pass through the diaphragm in the direction of the magnet housing. To prevent petrol from actually getting so far, a drain hole is provided just below the flange at the lowest point of the magnet body, and fuel emerging from this indicates almost certainly that the diaphragm is faulty, assuming, of course, that the flange bolts are tight. Regarding this latter point, leakage from the flange should not be confused with dripping from the drain-hole. In any case the former is very rarely encountered.

The diaphragm is obtainable as a new replacement part. To remove a faulty diaphragm, the magnet body should be held in the left hand, tilted slightly downward with the contact-breaker lowest. Then rotate the diaphragm with an anti-clockwise screwing motion, when it will be found that the armature spring pushes the diaphragm away from the housing. As it comes away, the eleven brass rollers will be released which locate at the back of the diaphragm. (These support it while at the same time allowing the necessary reciprocating action.) The diaphragm and its spindle, which are replaced as a unit, will now be free. Note that the large-diameter coil spring on the spindle is always fitted with its larger diameter against the magnet housing.

To reassemble a new diaphragm, first fit the coil spring, positioned as above, into the housing. See also that the small impact washer is in position on the spindle. Then feed the spindle into the housing, and as it meets the threaded centre of the rocker assembly at the other end of the pump, screw it in clockwise. Continue screwing until the rocker will not throw over as shown on Fig. 5.6, but do not 'jam' the armature in any way. Next hold the pump with the rocker end downwards, and refit the eleven rollers into their recess; this is facilitated by carefully turning back the diaphragm edge. When the rollers are properly located, hold the pump horizontally and try the action. Push the diaphragm centre firmly 'in', and unscrew the diaphragm spindle, a little at a time between pushing and releasing, until the rocker can just throw over. The diaphragm spindle must now be turned anti-clockwise until the holes in its edge line up with

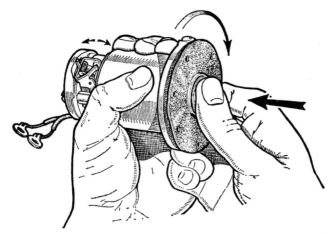

Fig. 5.6. Diaphragm is screwed 'inwards' until rocker ceases to throw over.

the flange holes; from this point, turn it in the same direction for four more holes, or two-thirds of a turn. In this position the diaphragm should be correctly set. The pump chamber body can then be refitted after cleaning and the flange screws inserted, but finger-tight only.

Stretching the diaphragm

At this stage, an assistant is very useful, but by no means essential. The pump earth terminal should be connected to the battery earth by any easily-arranged means. A lead should next be attached to the pump insulated terminal, and its other end left loose but available for contact with the other battery terminal. The 'drill' is now for the contact to be made, thus passing current through the magnet coil. This will pull the diaphragm to the full-suction position, $i\,e$, it will be stretched to its normal pumping extent.

The contact-breaker must, however, be prevented from operating and breaking the circuit at this point, and therefore either the points must be held together, or wedged together with a matchstick under one of the white rollers on the breaker mechanism. With the current 'on' in this manner, the six screws on the flange must then be tightened, tackling diametrically-opposite ones until all are tight. The battery can then be disconnected.

It will be obvious that an extra pair of hands is useful because of the number of operations, and the undesirability of having current passing through the coil for too long a period. At the same time, there is no need for panic if the job seems to be taking a long time as undue heating of the winding will show up long before any damage is done. With a battery on the bench and the pump held in the vice, it is quite easy to do the job single-handed, the contact-breaker points being wedged as described, and the circuit made with the left hand while the screws are tightened with the right. It should be borne in mind, however, that there must be no interruption of the circuit between tightening of the screws; tightening has to be done in one operation.

Testing the pump

It is possible that replacement on the car will be the only means of testing, though a much more useful rig can be arranged with a jar of petrol and rubber tubes on the inlet and delivery unions of the pump, and this enables the action to be checked more satisfactorily. The pump should, of course, be held in the vice, in its normal horizontal position. A good-sized jam-jar will serve as the fuel container, and the suction tube should reach to the bottom of this. The delivery pipe should be arranged so that it will deliver into the jar above the fuel level. Take care to use plenty of tubing so as to keep the petrol away from the pump terminals and contact-breaker.

When battery contact is made with the pump terminals, the pump should 'draw' quite quickly and should then proceed to pump steadily with a full, long delivery stroke. There may be a few air-bubbles for the first strokes, as the pump gets rid of air in its chamber and pipes, but these should not persist.

The pump is functioning correctly when it performs as long a stroke as possible, while at the same time allowing the contact-breaker to flick-over. If it operates unduly fast, it will still work, but will pump a reduced volume on each stroke, which means unnecessary wear and tear on the points and diaphragm.

If the pump appears to be working unduly fast and not using its

FUEL SUPPLY APPARATUS 97

full stroke before the contact-breaker flicks over, the remedy is to screw in the armature a little further, say one-sixth of a turn at a time. To do this it is, of course, necessary to remove the flange screws and, after making the adjustment, stretch the diaphragm as already described.

If, on the other hand, the pump hesitates, making a few strokes and then stopping and requiring a tap of the hand to get it going, it is probable, providing there is no obstruction in the pipes or valves, that the solenoid is unable to pull the armature far enough to allow the contact-breaker to operate; in other words, the armature must be unscrewed in the opposite direction to the above. If the adjustment has been made as described, however, everything should be satisfactory.

The contact-breaker assembly

The contact-breaker very rarely needs attention, the mechanism functioning faultlessly for very long periods. It will, however, be useful to detail the work involved in dismantling the components, in case undue exposure to dirt, or other causes, affect the operation.

As already mentioned, the bakelite cover is retained by a single fixing stud which also forms the live terminal, and carries a nut holding the cover and accessible after the terminal screw proper has been taken off. The cover sits on a further nut, which has also to be removed together with its lead washer (see Fig. 5.7). The

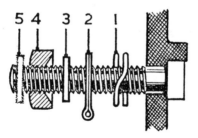

Fig. 5.7. Order of assembly of components on pump terminal.
1 Spring washer. 2 Cable tag. 3 Flat washer. 4 Nut. 5 Cover seal washer.

outer contact blade is held by a single screw with 5 BA thread and can be removed together with the screw and spring washer.

The bakelite contact-breaker pedestal is retained by two long screws with 2 BA threads, and these should be unscrewed and removed, with their spring washers. The whole breaker pedestal can now be taken off, the only obstacle being the lead from the magnet coil, which must be 'wangled' over its terminal without breaking it off the coil. This is merely a matter of using care and not tugging at the connection.

It will be seen that the rocker unit itself pivots on a hinge-pin which is not retained in any way except by the bakelite cover (when the latter is on). This pin can thus be pushed out from one end, and the assembly is completely dismantled.

The parts should be thoroughly cleaned, the contact points being treated in a manner similar to that employed for ignition points, *i e*, dressing up with a very fine file, and finishing off with a carborundum slip as shown on Fig. 5.8. The main thing about reassembling

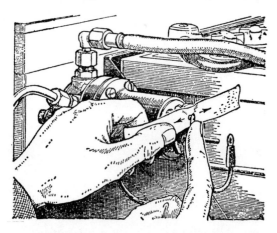

Fig. 5.8. Method of dressing points with very fine carborundum slip.

the rockers is that they must be perfectly free but with no excessive side play which would allow the points to misalign with each other. The hinge-pin is made of hardened material, and should always be replaced by the correct item; ordinary wire is no use whatever, except as an emergency measure if the pin is lost.

The side play on the rockers can be got just right by attention to the outer rocker, using a pair of thin-nosed pliers to square this up. When replacing the spring-blade carrying the other contact, note that the blade itself bears directly against the bakelite, the terminal tag coming on the outer side of the blade. Further, the blade will rest, if properly fitted, against the ledge formed below the opening in the bakelite through which the points operate.

When replacing the spring blade carrying the other contact, ensure the blade conforms to the position shown on Figs. 5.3 and 5.4.

Adjusting the points

Adjustment of the points may occasionally be necessary (as already dealt with) and, when correct, it will be found that the rocker will come right forward to the end of its travel with the points in contact, making their first 'touch' when the rocker has moved about halfway. If the spring contact blade point seems to be pressing too hard against the point carried on the rocker, it can be bent slightly so as to correct. Too heavy a pressure will prevent the pump from performing the full stroke and the rocker-arm from flicking-over, while too light a pressure means insufficient contact and liability to stop working at inconvenient intervals. When the pump has been completely dismantled, it is essential to renew all sealing washers, particularly those on the valve assemblies.

When refitting the pump to the vehicle, the earth terminal wire must not be omitted. The pump may run without it, but this is no criterion—it will run better with it. Pipes should be lined up carefully with their unions, and the nuts tightened firmly but not excessively. If all parts are clean, there will be no leaks.

The P D electric pump

Another type of electric pump, the P D, which is again of S U manufacture, is specially designed for mounting in the luggage boot adjacent to the fuel tank as shown on Fig. 5.9. Unlike the other types described, its construction is such as to discourage owner-overhaul, as the main units cannot be dismantled. The pump is mounted vertically, and comprises a main body with top and bottom covers housing the contact-breaker and fuel filter. The body contains the magnetic solenoid mechanism, which actuates the pump diaphragm below it by means of an oil-filled connection. This latter feature enables a very thin Terylene diaphragm to be employed, requiring little flexure. The body comprises a brass tube, inside which is a permanent magnet with steel polepieces, a plunger separated from the magnet by an insulated distance piece, and a spring to urge the magnet and plunger downwards. The insulated coil is mounted above this, and under the contact-breaker plate. The tube is filled with a light mineral oil and hermetically sealed, consequently the pump cannot be dismantled and in the event of trouble it must be exchanged for a service unit which will most likely be of the S P or A U F type. Details are shown on Fig. 5.10.

Fig. 5.9. Type P D pump as installed in Mini-Minor. Filter construction is shown inset.

Filter cleaning

The bottom cover is retained by a nut, and when the latter is removed the filter gauze is exposed. Blockage of the filter is indicated by a tendency to fuel starvation at high speed and load. The filter must be washed in petrol and blown through, at the same time cleaning the sediment from the surrounding parts and bakelite moulding. Whenever the cover is disturbed, a new cork gasket must be fitted; the washers on the bottom screw must also be replaced correctly.

The contact points are exposed by removing the top cover. It is important to note that there is no adjustment for the points, and that cleaning must be confined to drawing a piece of thin card between them while closed. When doing this, take care not to bend the blades. To check the action, the operating rocker can be moved by hand until its upper extremity touches the centre tube, the contact blades then bearing downwards on the bakelite platforms on the mounting plate. Positioned in this way, the blades should be parallel and straight, with a perceptible gap between the two pairs of points. On the rocker being released both blades should deflect upwards, the points firmly contacting. The gap between the extrem-

FUEL SUPPLY APPARATUS

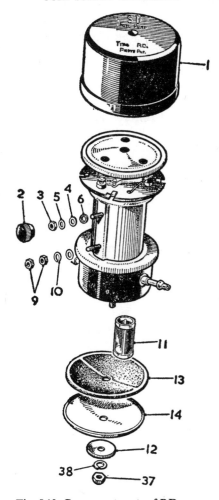

Fig. 5.10. Component parts of P D pump.
1 Top cover. 2 Knob for terminal. 3 Nut for terminal. 4 Plain washer. 5 Lead washer. 6 Insulating collar. 9 Nut for earth terminal. 10 Washer for earth terminal. 11 Filter. 12 Dished washer. 13. Cork gasket. 14 Cover-plate. 37 Nut. 38 Spring washer.

ity of the upper blade and its stop should be not less than 0.015 in. If through some misfortune misalignment has occurred, it may be possible to reset the blades by careful bending with a pair of thin-nosed pliers. Great care is needed, as it is essential that the downward-thrust of the blades on their stops is quite light, but at the same time positive.

Faulty action

The pump is unusual in that (unlike other S U pump types) it continues to 'click' when switched-on, even if the carburetter is full and there is no fuel supply defect. However, excessively rapid operation, with lessened fuel delivery, indicates an air-leak on the suction side. This may be checked by uncoupling the delivery pipe at the carburetter and leading it into a jar with the pipe-end submerged in petrol. When the pump is working, no air bubbles should be present in the flow from the pipe-end. If such are indicated, examine the gaskets at the filter end and make sure that the cover has been properly assembled. The fuel suction pipe should also be examined, particularly at its connections, and faulty components renewed if necessary.

The A C mechanical pump

This type of pump is used extensively, but being engine-driven it is not possible to pump up a full supply of petrol to the carburetter prior to using the starter, except by operation of the hand-priming lever attached to the pump body. Thus, engines equipped with this pump may require to be turned over for a few extra revolutions, compared with electric-pump-equipped vehicles, before the engine fires. The pump operates on the diaphragm principle and in this respect it resembles the S U. It is almost invariably mounted on the side of the engine crankcase adjacent to the camshaft and is operated by an eccentric mounted threreon. A flange and set-screw fitting holds the pump body to the engine.

The two sizes of pump most used on car engines are similar, but have different filter arrangements, the larger type having a glass filter bowl instead of a plain cover.

A section through the smaller pump is shown on Fig. 5.11 from which it will be seen that the pump diaphragm is operated by a vertical pull-rod integral with it, the pull-rod projecting below the diaphragm. Around the rod is a large-diameter coil spring which performs the delivery stroke. For the suction stroke, the pull-rod is pulled down by means of a tappet-lever operated off the camshaft. This tappet-lever is actually in two pieces which abut one against the other, and the portion of the lever which bears on the cam is spring-loaded inside the pump body, so as to keep the lever continually in contact with the cam even when the pump is not delivering due to sufficient fuel being in the carburetter.

In such circumstances it will be understood that the diaphragm will be held at its lowermost point, with the pump chamber full of fuel and the diaphragm spring exerting pressure thereon, ready for delivery to take place when required. The pull-rod, and its corresponding short portion of tappet-lever, will also be at their lowest

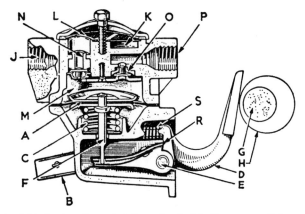

Fig. 5.11. Section through A C mechanical fuel pump, small type.
A Diaphragm. B Priming lever. C Diaphragm spring. D Rocker arm. E Pivot. F Pull-rod. G Camshaft. H Eccentric. J Fuel inlet. K Well. L Filter gauze. M Pumping chamber. N Inlet valve. O Delivery valve. P Fuel outlet. R Rocker abutment point. S Rocker return spring.

positions, and the other part of the lever will reciprocate idly against its loading spring, as the camshaft rotates.

Some pumps have a hand-priming lever projecting from the pump body. This is useful for test purposes, but normally should hardly ever need to be operated (certainly not on the road), if the pump is in good condition. If the engine comes to rest with the cam at the end of the suction stroke, this means that the pull-rod is at its lowest point, and obviously movement of the hand-priming lever will have no effect, as the diaphragm is already at 'full suction'. It is then necessary to turn the engine for a complete revolution, or half a revolution of the camshaft, so as to let the pump go to the 'end-of-delivery' stroke, when it will be found that the priming lever will operate the diaphragm effectively.

So much for the actual operation of the pumping element. The valves are situated in the underside of the top casting, and vary slightly in assembly according to the type of pump, though all are of the simple plate-type and readily removable. The diaphragm is quite readily replaced and impending trouble in this component is shown by a tendency for the pump to starve, particularly when faced by an extra 'lift', as when the car is climbing a hill with not much fuel in the tank and plenty of throttle. In such circumstances, even a minute puncture in the diaphragm will do nothing to help.

A rough check on the pump is to operate it by the hand priming lever after the fuel delivery pipe has been uncoupled at the carburetter end (if no primer is fitted, the engine must be turned by hand). A hearty spurt of petrol, free from air bubbles, should be delivered at each stroke of the pump.

Overhauling

The pump is not difficult to inspect completely, and for this purpose is readily removed from the engine by unscrewing its retaining nuts (or bolts) and the fuel pipe unions. It must be then thoroughly cleaned externally and the circumferential flanges, above and below the diaphragm, marked with paint to ensure correct reassembly in relation to each other. The screws around the flange may then be taken out, and the two flanges parted, which will expose the diaphragm. Its edge should be carefully separated from the bottom flange should it tend to adhere to this. The diaphragm is then turned through 90 degrees, which has the effect of unlocking its pull-rod from the actuating arm; the direction of turning is immaterial, this being of course in relation to the bottom pump casting. The diaphragm, complete with its rod, can then be lifted off. The valves and cages can be removed by taking out the retaining screws. The top cover, with filter gauze below, has a simple one-bolt fixing which varies with pump type. As far as the diaphragm actuating mechanism is concerned, it is unlikely that this will require any attention, and dismantling is not recommended, any undue wear being rectified by fitting an exchange pump. Components of the larger type of pump are shown on Fig. 5.12.

When the filter is removed, the fuel inlet is clearly visible. The passage of fuel is actually from this inlet, up through the gauze into the dome, and down the hole opposite the inlet to the suction valve. Below this is a small cavity which acts as a sediment trap for dirt which is centrifuged off the fuel flow as it turns upwards through the

FUEL SUPPLY APPARATUS 105

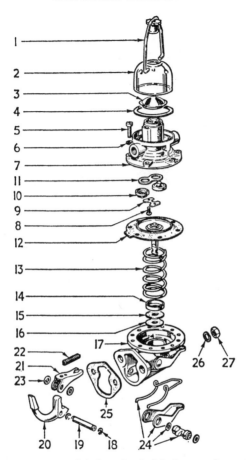

Fig. 5.12. Components of A C mechanical fuel pump, larger type with glass filter bowl.

1 Stirrup. 2 Sediment bowl. 3 Filter gauze. 4 Joint. 5 Screw. 6 Spring washer. 7 Body. 8 Screw. 9 Retainer. 10 Valves. 11 Upper retainer. 12 Diaphragm assembly. 13 Spring. 14 Cup. 15 Washer. 16 Washer. 17 Lower body. 18 Circlip. 19 Spindle. 20 Operating lever. 21 Operating fork. 22 Return spring. 23 Distance washer. 24 Priming lever assembly. 25 Gasket. 26 Spring washer. 27 Nut.

filter. This sediment trap can be drained occasionally by removing the small screw which will be found just below the suction union, on the outside of the pump body, but in addition to draining, it is advisable to help matters by probing with a match-stem to dislodge any hard dirt.

The valves and components should be washed in petrol. Re-assembly is in the following order.

Delivery valve-spring retainer, valve plate gasket, delivery valve and spring, inlet valve (on its seat), spring in disc centre, valve plate, retaining screws. Make sure that the inlet valve spring is properly seated on the valve plate. Some pumps have separate valve units, the inlet and delivery valves having their springs positioned below and above the valve-plate respectively. In this case the retaining plate has a spectacle-shaped gasket.

The filter gauze should also be cleaned in petrol, examined for microscopic punctures and a new one fitted if at all doubtful.

Diaphragm refitting

The new diaphragm is fitted by inserting its pull-rod downwards into the pump body, and turning the diaphragm until the tab on its edge points to '11 o'clock'. Press downwards on the diaphragm centre to compress the spring, at the same time turning it to the left so that the slots in the rod engage the bottom link. Finally turn a full quarter-turn, again to the left, and the pull-rod will be properly seated in its working position, with the screw-holes at the diaphragm edge properly lined up with those in the casting. Note that the tab will now be at '8 o'clock'.

The top casting can now be positioned by the paint-marks made before dismantling, and the screws inserted. The flanges must be scrupulously clean, and no jointing compound should be used. The screws are next tightened but only so that they barely compress the lock-washers. Now push the rocker-arm towards the pump, so as to pull the diaphragm down to the full-suction position. While holding the arm like this, tighten the flange screws a little at a time diametrically across the pump so as to obtain even pressure all round. The rocker-arm must be held until the screws are fully tight, and it will be noted that the diaphragm edge is just flush with the flange edges.

Fuel connections

The dome top, as mentioned before is held by one screw, and seats on a circular washer recessed in the top of the body. This washer should always be in good condition, as otherwise there will be serious air leaks, and the pump will not function properly. If the dome is tightened down too hard in an attempt to get a tight joint, the cover may be distorted, or the thread stripped. As filter cleaning is only necessary, assuming reasonably clean fuel, at about 10,000-mile intervals, it is good practice to renew the dome washer every time the dome is removed for this purpose.

Care should be taken that there are no leaks at the suction and delivery unions. Excessive tightening of these may lead to damaged threads, so that any leaks should be examined for cause.

Finally, do not keep a faulty pump in operation by using the hand priming lever. This should only need using for test purposes.

CHAPTER 6

The Lubrication System

Engine lubrication—Oil pressure—Other lessons from the gauge—Filters—Low pressure—Other causes of low pressure—Bearing wear—Auxiliary bearings—Choice of lubricants—Thin oil advantages—Draining the sump—Under-sump drain plugs—Flushing the sump—Renewable-element fiilters—Care in assembly—Throw-away filters—Fitting a by-pass filter—The connecting pipes

So far we have dealt with what might be called the routine maintenance of items which exert considerable influence on the good behaviour of the engine—the carburetter and fuel supply system, the ignition equipment, and the valve adjustment. Obviously, however, the correct mechanical functioning of the power unit depends on many other factors which do not require periodic attention. A proper understanding of these will nevertheless help not only to get the best out of the vehicle, but will also influence the peace of mind of the driver.

Engine lubrication

The operation of an engine's oiling system, nowadays completely automatic and almost 100 per cent reliable, is still something of a mystery to many drivers; even the enthusiasts occasionally find it hard to reply to some queries, usually connected with pressure. It will be useful, therefore, to consider for a start the general operation, and then decide the best way of ensuring that the maker's designed method of fool-proof lubrication operates to the best advantage.

Modern engines display a considerable degree of unanimity in the design of the system, in that the oil supply is carried in a chamber under the crankcase, a pressure pump delivering lubricant to all moving parts by way of various ducts and pipes, after which the oil drains by gravity back to the base chamber. As shown diagramatically on Fig. 6.1 this layout seems quite simple, but there are several points to observe.

The oil-pump itself may be mounted down in the base chamber or externally, but is almost invariably driven from the camshaft itself or by an extension of the drive thereto. It is of a type which can usually be relied on to prime itself; that is, it will 'pick-up' oil, in

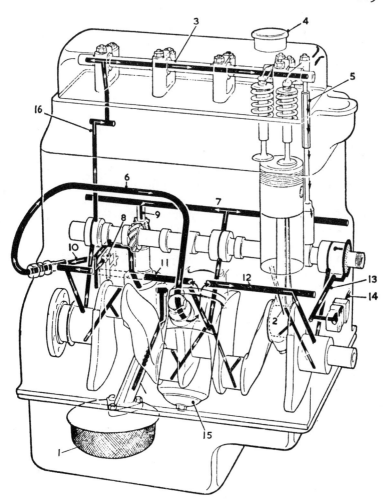

Fig. 6.1. Lubrication system of Morris-Oxford engine shown diagrammatically.

1 Oil strainer. 2 Oil squirt to cylinder walls. 3 Rocker feed hole. 4 Oil filler. 5 Oil return to sump. 6 External pipe. 7 Low pressure gallery. 8 Relief valve. 9 Feed to gears. 10 Main feed. 11 Oil pump. 12 Main high pressure gallery. 13 Feed to chain tensioner. 14 Oil bleed to chain. 15 Full-flow filter. 16 Feed to rocker gear (intermittent).

most cases (even when the suction pipe or intake has been completely emptied), as soon as a supply is restored.

It also delivers at a considerable pressure, and the quantity of oil delivered varies just about in direct proportions to engine speed; that is, if the pump delivers two gallons per minute at 2,000 rev/min, it will deliver four gallons at 4,000 rev/min.

Now, obviously, the pressure developed by the pump is used to force the oil through clearances. In fact, the back-pressure put up by these clearances determines how much pressure is provided at the pump outlet. If the outlet is perfectly open, as, for example, when disconnected from its distribution system, negligible pressure will be present, though a large quantity of oil will be delivered.

The clearances, which may comprise the running 'fit' of the bearings, and the restrictions provided by specially drilled holes, in certain cases, are, of course, fixed by the engine design, and only vary as a result of wear or damage. That being the case, it will be evident that, as the quantity of oil delivered by the pump is dependent on its speed, the pressure at the pump delivery will also vary in proportion. In practice, this does not happen, as is evident from the pressure gauge.

Oil pressure

At any engine speed above idling, the pump is designed to provide a steady pressure at all speeds. This is done, in the first place, by making the pump amply large enough to deliver more than sufficient oil for any conditions of running, so that there is at all times a surplus delivered. This surplus is by-passed from the delivery side of the pump, back to the sump again, by way of a spring-loaded relief valve. The load on the spring of this valve determines how much oil is by-passed, and consequently the pressure of delivery.

A knowledge of all this is useful in reading the pressure gauge. The pump delivery outlet is usually connected to an oil distribution gallery running along the crankcase and cast therein. From this gallery, oilways lead to the various bearings. With this arrangement it will be evident that the first obstruction to free passage of the oil is the bearing clearance. From this, it follows that, if the bearing clearances, or the oilways thereto, are closed completely, so that no oil is flowing at all, the oil in the gallery will be at maximum pressure, and this pressure will be registered on the gauge. The only oil flow taking place will be through the pump relief valve, back to the sump.

In practice, of course, conditions could hardly arise whereby no oil whatever was able to flow to the bearings from the gallery. In very cold weather, however, something approaching this is by no means uncommon. The thick, congealed lubricant flows reasonably freely into the large dimensions of the gallery, and, of course, registers a reassuringly high pressure on the gauge. A woefully small quantity is forced into the bearings, and the flow does not become adequate until such time as the oil has thinned out sufficiently to move more freely. The bulk, of course, by-passes via the relief valve. In these days of high-grade thin oils, the foregoing conditions do not arise to anything like the same extent as was common not very long ago. Modern oils, even when very cold, will flow fairly freely between the relatively large clearances allowed by present-day methods of bearing design and construction. Nevertheless, it is obvious that the showing of a high oil pressure is no guarantee that the lubricant is actually going where it should in sufficient bulk, but fortunately the gauge provides another sign which is valuable to the discerning driver.

Although the by-pass valve fixes the oil pressure (in conjunction, of course, with the bearing and the other clearances), this pressure is not absolutely steady, since, naturally, cold oil flows with less freedom than warm oil in all parts of the system, and this includes the relief valve. For this reason the oil pressure as registered by gauge will usually be a little higher when the engine is stone-cold than at normal running temperature, and a study of the position of the gauge pointer under various weather and running conditions will enable the driver to judge when the oil has thinned sufficiently to permit the engine to be 'worked'.

If a fine-gauze suction filter is fitted to the pump inlet, the 'cold' pressure may take considerable time to build up. In these circumstances also, driving must be adjusted accordingly.

This information 'per gauge' is extremely valuable. Naturally, however, this variation of pressure will persist for quite a long time, and in fact, on some short 'ride-to-work' journeys, the oil may never thin out to the same extent as it would under normal touring conditions. Using modern lubricants, however, there is no need to be panicky about 'using' the engine just because the pressure is a little different from normal, so long as reasonable discretion is exercised. As soon as full pressure is registered on the gauge, after the initial start, a couple of minutes at a fast idling speed will ensure that the oil gets moving, and after that, there will be no harm in using up to half the maximum rev/min right away, on an intermittent basis—which will probably be inevitable in traffic anyway.

On the other hand, the rapid warming-up technique should not be overdone. The writer has come across owners who think nothing of screaming their engines up to astronomical revolutions, straight away, before moving at all, usually because the carburetter is so hopelessly adjusted that a get-away from cold involves either a series of starts, stops and jerks, or excessive use of the choke. It cannot be too strongly emphasised that excessive engine speed in this manner will cause actual metallic contact between bearing surfaces; in fact, this contact is frequently audible under such conditions.

Other lessons from the gauge

It will be evident from our description of the pump layout that in modern designs, the output volume from the pump is more than sufficient to maintain pressure throughout the engine speed range. Thus, the pressure should remain constant, once the oil has reached its working temperature. The only exception is at very low idling speeds, when a drop in pressure is to be expected. Even when idling, however, an adequate pressure must be shown; for a normal pressure of 50 lb an idling pressure of about 12 should be right. The maker's handbook sometimes gives more specific information on such points. Obviously, there is no harm, in fact it is a sign of good condition of bearings, if the idling pressure is not far below the normal pressure.

The running pressure may vary due to road conditions. For example, during and after a shower the pressure may increase slightly because of the sudden cooling of the engine sump. After a bout of very fast driving, or hill-climbing in the lower ratios, a slight drop in pressure may take place. Such changes should amount to very little, if the engine is in good condition, and any great variation must be suspect.

Filters

Oil filters may be provided on the suction or delivery side of the pump, or on both. The suction filter usually comprises a gauze screen, which can be seen on Fig. 6.1. This is designed only for preventing really large bodies from entering the pump; for example, metal pieces such as split-pin legs or machining swarf which might be capable of wrecking the pump. This gauze is only accessible after removing the sump and cleaning need only be done at major overhaul periods.

The filter on the delivery side is usually of the renewable-element pattern, designed for thorough filtration. The mileage at which the element must be renewed is stated by the car maker, but is in the region of 6,000 miles. As shown on Fig. 6.2 provision is made, in

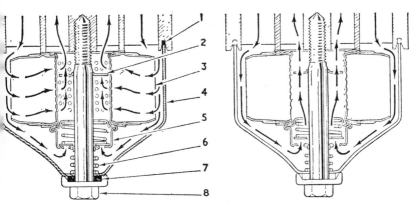

Fig. 6.2. Section of full-flow filter with single-bolt fixing on Triumph Vitesse engine. Flow with blocked element is shown on right.
1 Gasket. 2 Baffle. 3 Filter element. 4 Casing. 5 Spring housing. 6 Main spring. 7 Washer. 8 Main bolt.

the filter itself, to bypass the oil, so that even if the element becomes completely choked and incapable of passing oil through it, unfiltered oil can reach the engine. This guards against engine damage even with gross neglect of the filter, but is obviously a state of affairs which should never be allowed to happen.

A third type of filter is the by-pass type, which is connected in a subsidiary circuit of its own, between the main oil gallery and the sump. None of the main oil supply to the bearings from the pump passes through this filter. The subsidiary circuit is of small capacity and the filter has a very fine element. Thus, a little of the oil is circulating through the filter all the time without influencing the main oil supply. Eventually, of course, all the sump oil will have passed through the filter.

The by-pass type of filter is very effective. As it is not dealing with a large volume of lubricant, and as the supply to the bearings is not dependent on a free passage through it, the effectiveness of the filtering medium can be the first consideration in design of the component. The small quantity of oil flowing can thus be thoroughly cleaned. By-pass filters can be fitted to any engine as an adjunct to the existing system, and are a very sound investment.

Low pressure

An uncalled-for drop in oil pressure is at all times a depressing occurrence. If this happens suddenly the usual reason is shortage of oil. It is a surprising fact that even today there are drivers who think that there is some connection between the quantity of oil in the sump and the gauge pressure, and that as the sump level falls, the poundage on the gauge will conveniently drop progressively to so inform them. That is, of course, not so. The pump will draw, and deliver at full pressure, so long as its intake is submerged.

As soon as the oil level falls below the intake, and the pump draws air, the pressure will fail completely. The difference between these two conditions may be only half an inch or so of oily sludge in the bottom of the sump. On some designs, impending oil shortage is shown by a sudden drop in pressure when cornering, as under these conditions the oil may surge away from the pump intake. The pressure is then restored when the level becomes steady again. Operation under such conditions is bad for the bearings, particularly as such temporary starvation may well occur at very high speeds.

This type of starvation can be guarded against by keeping an adequate level of lubricant in the sump at all times. It is very bad practice to allow the level to fall below the half-way mark on the dip-stick, and not only because of the possibility of starvation. It will be obvious that the less oil there is in circulation, the more heat will be given to it, and although dyed-in-the-wool theorists may aver that ultimately the heat is radiated from the exterior of the sump, etc., and that the oil merely acts as a transferring agent, the fact remains that with the requisite quantity of oil in circulation a much happier and reliable engine will result. The reason for this is that the heat transfer takes time and, in practice, a small quantity of oil becomes much hotter than a larger quantity. This can be seen quite readily on cars fitted with a thermometer to measure oil temperature.

Other causes of low pressure

Apart from shortage, or complete lack, of oil, the usual cause of low pressure is, quite simply, bearing wear. It is not unknown to try to reassure owners regarding a decrease in pressure, by suggesting such items as worn pump internals, stuck relief valve or weak springs on the same, and so on. Such happenings are rare, whereas a progressive decrease in pressure over a long mileage, as a result of the bearing clearances increasing, is inevitable.

We have already explained that the developed pressure in the oil distribution gallery depends on these clearances. The degree of pressure drop caused by wear at the clearances will thus depend on the pump volume. If this is large, pressure will still be maintained even though a larger quantity of oil will pass through the bearings. In fact, so long as the pump by-pass valve operates at all, its spring will determine the pressure. When the time comes that all the oil is passed through the oilways and none through the by-pass valve, a drop in pressure will be the result.

This will help to explain a popular fallacy that oil pressure can be restored by increasing the loading on the by-pass valve spring. Consider an engine with well-worn bearings. When the oil is cold, some of it will pass through the by-pass, even though more than the designed quantity will be able to go via the bearing clearances, so that the pressure will be somewhat below normal. If now the by-pass valve spring is packed up, or replaced by a stronger one, the valve will be unable to divert so much of the supply so that more will be forced through the bearings. In this way, the 'cold' oil pressure can be made to appear normal. As soon as the oil thins out, however, it will pass quite freely through the clearances, none at all going via the valve, and down will come the pressure. No attention whatever to the relief valve will restore the 'hot' reading, and this is the one that matters.

From the above it will be evident that a very wide variation in oil pressure between hot and cold must give rise to suspicion. It is almost invariably indicative of bearing wear, and nothing short of bearing renovation will put matters right. Thus, it is a wise precaution when vetting a used engine, to run it until the oil has become thoroughly warm, and to watch the pressure under these conditions.

Unless the oil pump has been dismantled and incorrectly reassembled, a pressure drop due to a fault therein is most exceptional. The same applies to the by-pass valve, though in this case it is just possible for a speck of grit to prevent the valve from seating properly, thus reducing the pressure. It will be obvious, however, that the valve spring is very strong, relative to the size of the valve, and the oil flow is normally sufficient to keep the passage quite clear.

Bearing wear

When the main bearings wear, the extra flow of oil through their clearances will normally pass straight back to the sump. In very exceptional cases, it may be found that the oil-return arrangements on the crankshaft, particularly at the flywheel and clutch end of the shaft, are unable to cope with the extra flow and a slight leakage of oil into the clutch housing, and thence on to the garage floor, will result. As regards big-end bearing wear, the larger quantity of oil escaping from these bearings will, of course, augment that normally flung up into the cylinder bores, and the extra oil on the cylinder walls may prove incapable of being controlled by the oil-control rings on the pistons. In such cases a smoky exhaust and oiling of plugs will be the evidence.

When it is not possible for the engine to be renovated just at the time that items such as the foregoing suggest that overhaul is desirable, a temporary restoration of pressure, and mitigation of oil consumption, can be effected by substituting a heavier grade of lubricant for the one recommended. Even when hot, the thicker oil will be more resistant to passage through the larger clearances, and thus a pressure of something like normal will be maintained. This substitution of a heavier oil will serve its purpose for only a limited period, since once bearing wear really starts, the extra relative movement between shafts and bearing shells encourages progressive and rapid wear. Further, the use of thick oil is disadvantageous as it increases the frictional losses in the engine (power being absorbed in the oil film), and great care has to be taken when starting from cold, in view of the heavy load on the pump, and the sluggishness of the flow to the bearings.

Sufficient has been written to show that the oil gauge is a most useful accessory from which much can be learnt regarding the engine's well-being. The question might be asked as to what guidance can be obtained from the telltale light which is fitted to a great many cars instead of a gauge. This light only glows when there is negligible pressure in the system, say 4 to 6 lb per sq in. The most satisfactory answer is to fit one of the proprietary gauges which are available, and which are extremely simple to install. If this is not acceptable, the best compromise is to carry out the recommendations set out, just as if a gauge was fitted.

We have so far indicated that wear in the 'big' bearings of the engine, *i e*, those contained in the crankcase, is the usual cause of a normal and progressive drop in oil pressure as the miles accumulate. This is, in fact, the case. It will, however, be appreciated that from the main oil gallery, an external pipe, or a special oilway, is required to conduct lubricant to overhead-valve gear components, in engines of o h v type, and an unduly free passage here will cause a drop in pressure throughout the system.

Auxiliary bearings

In the case of push-rod-and-rocker o h v gear, the oil is usually fed, by one of the means stated, into the rocker shaft, which is hollow. It then proceeds via drilled passages to such parts of the rockers and their associated bearings as the designer considers advisable. The size of the passages determines to a very great extent how much oil passes, and thus the pressure built up, but wear in the rocker bushes will, of course, tend to allow oil to escape from their ends. These rocker bushes usually wear very slowly, providing they are of the specified material, and their rate of wear should be much less than that of the 'downstairs' bearings already dealt with, so that pressure drop as a result of wear in the o h v department is unlikely.

In the case of overhead-camshaft engines, it is sometimes necessary to allow for a free passage of oil to certain components such as driving gears. In such cases, the main supply is frequently restricted by means of metering plugs inserted in the oil passages. Thus, the supply has to pass through that restriction. Low pressures in such engines have been known to occur through tampering with the restrictor plugs.

To sum up the conditions necessary for efficient engine lubrication, we can conclude that the designed system is to all intents and purposes 100 per cent reliable. To ensure its operation in that state, it is up to the driver to keep the filters clean, which nowadays generally means renewing the 'thow-away' filter element at the stipulated intervals. To change the engine oil regularly, as specified by the manufacturers of the engine, and not to believe that this recommendation is a ramp encouraged by the oil concerns. To use the proper grade of oil, and not one thicker because it 'looks safer'. To keep the engine exterior clean, so that any slight leaks, as at the fiter cover, external pipes, flange joints, and so on, can be spotted before they become serious. To maintain the correct oil level on the dip-stick. And, finally, to keep the exterior of the sump clean so as to encourage heat dissipation.

Choice of lubricants

Modern oils are much thinner than of yore, and many old-timers are apt to look askance at them. It is, however, a fact that whilst the modern varieties flow freely when cold, they also retain their property of viscosity, or oiliness, when really hot, which the old, thick lubricants could not be relied on to do at all times. There was a time when hardly any brand of mineral oil was considered absolutely safe under extremely arduous racing conditions, and castor-oil-based lubricants were the prime favourites, as exemplified by the famous Castrol 'R', the tang of which was a joy to the enthusiast. Nowadays, of course, thin mineral oils are regularly used for Grand Prix racing with complete reliability, and have resulted in much cleaner engines both internally and externally; not a small point when rapid overhauls are required.

Most manufacturers of modern cars are tending to specify thinner oils, and whereas a viscosity equivalent to the measure known as SAE 40 or 50 was usual many years ago (the higher the number the 'thicker' the oil), present day designs run quite happily on SAE 30 or even lower. Furthermore whereas the transition from summer to winter used to indicate a change to thinner oil in the interests of easier starting and quicker cold flow, it is quite usual nowadays to use the same rating all the year round, at all events in the British Isles.

Thin oil advantages

It will be apparent that so long as the oil chosen can maintain an unruptured film between bearing surfaces under all conditions of heating, the thinner it is the better. Not only does it flow more easily from the cold condition, but it creates much less friction in bearings and between cylinders and pistons. In this way, more power is available at the engine flywheel, because less of the power produced on the pistons is wasted in shaving through a thick film of lubricant. It is probable that were it not for the inherent conservatism of the average motorist, and also of some engine makers, even thinner oils would now be in popular use. As it is, one or two makers specify SAE 20 for normal temperature conditions, while experiments with SAE 10 and even thinner grades have not shown any disadvantageous results.

Multigrade oils are another modern development which, in a way, enables one to get the best of both worlds. A typical viscosity rating is SAE 10W/30: this indicates that the characteristics are those of an SAE 10 oil when cold and of an SAE 30 oil when hot. Thus cold starting is assisted by fluidity and prompt circulation, yet there is plenty of 'body' to withstand fast motoring when hot.

The question of detergents caused much debate among keen

owners when oil companies began to incorporate them. There was no doubt that they kept a new engine clean, but anxiety was felt about the dislodging of the more-or-less harmless dirt which used to build up in an old engine: might it not block up oilways, cease deadening the sound and provoke leaks where none were before? Somehow, thanks to careful use and regular oil changes, engines survived the transition, and now the problem no longer exists, since virtually all oils have contained detergents for long enough to wash away the most stubborn deposits of sludge.

As regards additives intended to be put in the oil by the owner, the best advice is to ask both the car maker and the maker of the engine-oil for their opinion, and abide by this.

Draining the sump

Maintenance of the engine's lubrication system is one of those jobs which, while demanding attention of a simple nature, is all too frequently left to the service station. This is rather surprising, as it is possible to effect quite a reduction in running costs by doing the necessary at home. As an example, engine oil purchased in bulk, say 5 or 10 gallons at a time, is considerably less expensive than payment for merely the amount which the sump holds, which in the case of a medium-sized car is about 1 gallon. This payment forms, of course, the basis of the charge for draining and refilling at the local garage. A further advantage of having a substantial oil supply at home is the fact that the same brand of lubricant is in use at all times, while topping-up can be done conveniently without either departing from the selected brand or letting the oil level fall to a dangerously low point.

It will be obvious that the best time for sump draining is when the oil is really hot—after a good long run. Draining should be carried out at the stipulated intervals without fail. The permissible mileage between oil changes has tended to increase in recent years, thanks to the improved quality of lubricants and filtration arrangements.

It is undeniable that draining is one of the least attractive operations, in that it generally necessitates a degree of grovelling under the chassis. It is possible by a little preliminary work to diminish this side of the business, the idea being to make up one or two simple accessories.

The first of these is a suitable container for the drainings. Knowing how much the sump holds, a suitable bowl, such as a metal domestic kitchen type, should be obtained, provided with handles for carrying. Make sure that the bowl is shallow enough to go under the drain plug. Then obtain about 3 ft of iron or steel rod—low-quality material is quite suitable and cheap—and bend one end into a loop which encircles one of the handles of the bowl. Close the loop completely so that the rod cannot come off the handle. We now have a draining-tank complete with extension handle which can be pushed under the engine without undue exertion. Incidentally, this item will be found invaluable for other tasks such as radiator-draining.

The next accessory is a device for removing the sump drain plug. The plug is usually provided with an external head of hexagonal or square shape; a further variation is that the 'shape' may be either internal or external. Whatever the design, it is intended to be undone by application of a suitably-shaped tool, and our object is to furnish a tool which can be operated from a reasonable distance.

When the drain plug is on the side of the sump, as on Fig. 6.3, and

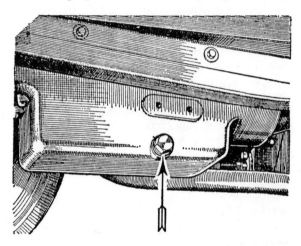

Fig. 6.3. Some engines have side-mounted sump drain plugs; early M G is shown.

providing nothing gets in the way, it is possible to make up a fitting using a tool or spanner of the type required, welded or otherwise secured to a long rod, something like a jack handle. In order to dodge obstructions, it is quite feasible to fit a simple 'universal joint' in the rod by means of a short length of tube and cross-pins, thus allowing the rod to deviate from the axis of the plug without coming off it.

Under-sump drain plugs

In cases where the plug is right under the sump the provision of an extension spanner is rather more difficult, as movement is naturally limited. If the plug has an external hexagon (and this type is often found in such positions), a ring spanner, or tube spanner, of the kind having twelve angles of attack, should be obtained, and a long handle fixed to this. Sufficient movement will be obtainable on each engagement to enable the plug to be slackened.

Coupled with the use of these aids to maintenance is one important proviso. The drain plug must not be too tight. The writer has come across cases where brute force, and the ruination of the plug, have been necessary to remove it, after tightening by someone armed with a hefty socket-spanner plus a long tommy-bar. Admittedly, the consequences of a lost drain plug might be serious, especially when no oil-gauge is fitted, but this is no excuse for excessive force, particularly when this involves such material as brass and aluminium, the threads of which are not difficult to mutilate irreparably. Any form of extension spanner must inevitably have a certain degree of 'spring' in it, and will thus be useless on a plug that has been put in too tight.

Of course it is necessary, before making up the accessory, to find out what type of plug is used. It should be possible to borrow a spare plug for this information, and this is preferable to obtaining the measurements from under the car, which is likely to be an unreliable method. Or, of course, there may be a spanner in the toolkit, clearly designated as being for the purpose, and this can form the basis of the extension tool.

So much for the possibilities of trouble-saving equipment. Thus armed, draining should not be too difficult. The plug should be slackened before the container is pushed under. When the plug is finally removed, with the container underneath, it can be allowed to drop into the latter, and subsequently fished out while the used oil is being poured away to waste. This is a much less messy procedure than attempting to whip the plug clear of the gush of hot oil as it emerges.

Although most of the oil will drain in a few minutes it is a good idea to leave the plug out for as long as convenient, to let the remaining drips get away. It is advisable not to 'motor' the engine round on the starter while draining is going on, as this will tend to empty the pump and oil gallery. In such a case, damage may result if the pump does not pick up the fresh oil immediately after refilling. It is agreed that a certain amount of 'old' oil will be contained in the lubrication system, and will mix with the new, but this is preferable to leaving the bearings without any at all, even for a short period.

Flushing the sump

On the question of flushing out the sump before putting in the fresh oil, there are various opinions. The writer inclines to the view that providing the oil is drained when it is really hot, all practical steps will have been taken to remove such impurities as are capable of being removed by draining, and no more will be got out by flushing. He also has a certain objection to introducing any substance other than lubricating oil into the sump. Under no circumstances must paraffin or similar liquid be used.

Before replacing the drain plug, make sure its washer is in good condition. If so, there is no need to over-tighten the plug, and it will be readily removable for future operations. The washer is usually of copper-asbestos, or sometimes of hard fibre. Washers of soft material are definitely barred, as shrinkage with temperature variation is liable to cause the plug to slacken off.

Tighten the plug quite firmly and decisively, but do not exert a superhuman effort to make sure it is tight; leave well alone. If you have a feeling about the plug coming out, in spite of all these assurances, have it drilled across its head diameter with a small hole, and thread a length of copper wire through, anchoring the wire fairly tightly to any convenient bolt-head on the engine. This precaution will prevent the plug from unscrewing very far, even if it should slacken-off slightly.

Apart from periodic changing of the engine oil regular attention to the oil filter is necessary, though with modern designs this may be only required at infrequent intervals, say up to 6,000 mile or so. The latest filter layouts make the task very easy, but there are many cars on the road which need rather more than a few moments' work to effect filter maintenance. It will be useful to describe one or two of the arrangements likely to be met with.

Renewable-element filters

This type of filter comprises a housing mounted on the side of the crankcase as shown on Fig. 6.4. Many modern engines have a machined facing on the crankcase side, this being formed with ducts for the oil so that there is no exterior piping. The complete filter is secured by a central bolt; the filter shown on Fig. 6.2 is of this type. The housing may be formed of a pressing or an alloy casting, with

THE LUBRICATION SYSTEM 123

provision for access to the interior which contains the filtering element. The oil passages are arranged so that the lubricant passes from the outside to the inside of the filtering medium, and the whole of the surface is used in filtering. The element is discarded at the

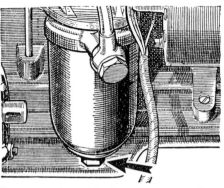

Fig. 6.4. Filter mounted on side of crankcase, on Riley 4/72 engine. The single bolt securing canister is shown arrowed.

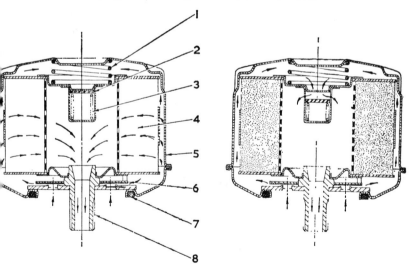

Fig. 6.5. Filter of 'throw-away' type with sealed canister, and screwed spigot for mounting (Triumph Herald). Flow with blocked element is shown on right.
1 Spring. 2 By-pass valve. 3 Valve spring. 4 Filter element. 5 Casing. 6 Base. 7 Gasket. 8 Outlet spigot.

specified intervals and replaced with a new one; in some cases it may be of a type capable of being cleaned in petrol, and in such cases cleaning may be required at 4,000 mile intervals and complete renewal at 12,000 miles or so. If the housing is of the vertical type with a removable cover for access to the element, it is important not to allow foreign matter to drop into the housing when the cover is off. Another type shown on Fig. 6.5 comprises a competely renewable unit attached to the engine by a central screwed spigot. Both the casing and element are replaced 'in one' when required, without disturbing any oil connections, the latter being formed in the attachment spigot.

A further type has a cylindrical housing secured by a central bolt to the top (which is bolted to the engine and contains the oil passages) so that the cylindrical housing complete with element can be dropped down. Thus, a very complete cleaning operation may be carried out on the inside of the housing, a new element inserted, and

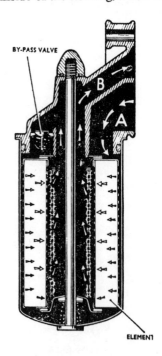

Fig. 6.6. Section through a renewable-element filter showing oil-flow and long canister bolt, on Hillman Imp engine.

the assembly completed, without any fear of foreign matter gaining access, as everything is done from below the cover. (See Fig. 6.6.)

All things considered, a passable way of tackling element-changing on a filter with a top-detachable cover, is to remove the cover, lift out the element, replace with a new one, and refit the cover. Any attempt to drain the housing of oil, or to clean it out with rag, is likely to do more harm than good. The ideal method, of course, and one which the really conscientious driver follows, is to take the complete housing off before dismantling any part of it. The cover can then be removed in comfort on the bench, everything cleaned thoroughly in petrol, and reassembled with the new element complete, before reattaching the unit to the engine.

Care in assembly

Whichever of the above methods is followed, there are certain things to be watched. The housing is sometimes of cast aluminium, and therefore easily distorted. The same applies to the cover.

Bearing in mind the oil pressure which exists inside, it will be obvious that the slightest leak will pump all the lubricant over the side in very quick time. Therefore, after the lid has been removed see that all traces of the old joint-washer have been got rid of, and that the mating surfaces are perfectly clean. If a new washer cannot be obtained, of the correct type, make one out of good quality brown paper. Before refitting the lid, see that all studs are clean and their threads in good order, otherwise a binding nut may give the impression that it is tight, before the lid has been pulled right down on its joint faces. Spring washers should be used under nuts. The washer should be anointed lightly with jointing compound on both its sides and placed over the studs. Then fit the lid (not forgetting to put the filter element in first) and replace the spring washers and nuts. The latter must then be tightened down very carefully and evenly, doing them up in diametrically opposite pairs, and tightening only a little at a time. A spanner of standard length should be used, the squeezing-out of jointing compound being a sign that the nuts are nearly tight enough. Then give each a final pull-up.

If the complete housing has been removed from the crankcase, this will involve remaking another joint, *i e*, the one between the supporting flanges, assuming these flanges form the oil connection. The same care is necessary as for the lid joint, and similar material can be used for the washer. It is, however, stressed that the maker's washers are preferable to home-made ones. If a washer has to be made for this joint, note that there will be oil-passage holes therein, as well as stud holes, and the flange on the filter body should be examined for their position. When refitting the housing over the studs, make sure that no dirt gets on the washer or flanges before bolting-up; there may be a lot of dirt around this area of the engine.

If the oil is taken to and from the body by external pipes instead of through the flanged joint, these pipes will, of course, need removing before the body can be taken off the engine. The pipes are usually connected to unions which in turn are screwed into the casting. When unscrewing the union nuts on the pipes, the second hexagon, on the body union, should be held by another spanner. Otherwise, use of a single spanner on the union nut only is liable to shift the whole union, and thus to twist the pipe beyond repair, instead of parting the joint at the union nut. If by mischance the union does become loose in the filter body, it can become an awkward joint to get tight, and a leak here is fatal. To effect a remedy, it should first of all be ascertained that the threads of the union are in really good condition, likewise, the threads in the casting. A well-fitting red fibre washer (no other kind will do) is required on the union. Jointing compound should then be applied both to the union threads which go into the body and to the fibre washer. Then, with the filter body held in the vice, tighten up the union with a well-fitting box spanner. Do not tighten half-heartedly—pull up really tight, say with leverage equivalent to a 5-in. spanner-shank and average strength of wrist. On any subsequent dismantling, use the two-spanner technique to ensure that the union is not again disturbed.

The foregoing does not of course apply in cases where the unions to receive the pipes are welded or brazed on to the filter housing, as there should be no leakage from these.

The main item when refitting the oil pipes is to make sure that they line up with their unions. It is no use getting them more or less in line, and then trying to pull them into position by tightening up the union nuts. This puts an excessive stress on the union threads and other parts, and due to the unequal seating pressure of the coned pipe-end on the union seat, an oil-tight joint is unlikely to be achieved.

The more usual type of renewable element filter used today, with a single-bolt fixing for the housing canister, is considerably simpler to deal with than the type just described. The housing containing the element is detachable without disturbing any oil connections, as removal of the long centre-bolt from the bottom enables the cylindrical housing to be slipped down complete with the element inside it.

THE LUBRICATION SYSTEM

The whole assembly can be cleaned at leisure, and a new element inserted, when it only remains to slide the housing and element into position and tighten up the securing nut from below.

The only points needing attention are the sealing washers. There is usually one at the top, between the housing rim and the shaped cover. This washer is often contained in a circular groove in the cover, and does not readily become displaced. Consequently, it should not need renewing as a matter of routine, but only if it shows signs of breakage when the housing is dropped away. A second washer, usually of fibre, is located under the bolt head. Naturally, the leakage area at this point is small, and there is no difficulty about obtaining an oil-tight joint.

'Throw-away' filters

So much for filters which have a renewable element. The other type, which has many advantages, is generally known as the 'throw-away' pattern, as the whole unit, container and all, is discarded when the specified mileage has been completed. This has the enormous advantage of making the ingress of foreign matter almost impossible when filter-changing. The early type shown on Fig. 6.7

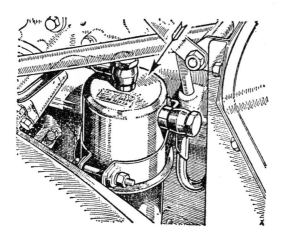

Fig. 6.7. Filter of 'throw-away' type with external pipes and retaining clamp, on early M G engine.

has external unions to which the inlet and outlet pipes are attached, these being closed by solid plugs when the filter is purchased. It can be mounted in its support on the engine and made secure, and last of all the plugs are removed and the oil pipes connected to the unions.

The method of support of this type of filter is usually by a band-type clamp with a simple form of securing nut and bolt, and a registertolocate on the cylindrical filter body. Removal and replacement thus presents no difficulty, and, so long as the aforementioned plugs are left in place on the new filter until the last minute, there is no necessity for a high degree of cleanliness in the surroundings. This is a boon when time is precious.

The 'throw-away' type filter may cost more than the renewable element only, but has the advantage of being entirely free from potential leakage, as it has no detachable cover, and there is no chance of foreign matter entering while assembling on the engine. Later designs have a single-point attachment similar to that used on renewable-element filters.

Filters of the by-pass type are made both as 'throw-aways' and with renewable elements, and the precautions to be taken do not differ from those already detailed for full-flow filters. It has already been remarked that the use of a by-pass filter on an engine already equipped with a full-flow type makes the lubrication system 100 per cent efficient. The fitting of such a filter does not present any particular difficulty, and filter makers provide standard fixing kits for most popular cars. If any comment be offered on these, it is merely to the effect that the kits are designed for the majority, and that the owner who likes to do a bit of thinking may quite easily find a method of mounting the filter on his particular engine which is an improvement on that laid down by the filter maker. The individual approach may, of course, involve the making-up of brackets and so on, of a more elaborate type than those normally provided by the filter maker.

Fitting a by-pass filter

As we have already mentioned, the by-pass filter is connected in a circuit of its own, between the pressure side of the pump and the oil sump. Some makers of cars provide a blanked-off tapped hole in the oil gallery itself to facilitate connecting the filter feed-line, by substitution of a screwed union of suitable size for the blanking plug. As regards the feed-back to sump, this can be connected to the overhead-valve rocker cover in o h v engines or to the valve-cover plate in side-valve engines. In both cases an alternative is to fit a union as low down as possible in the oil-filler neck (where this is provided), but if this neck also acts as a breather it may be necessary to provide an internal down-pipe inside the neck to ensure that crankcase pressure does not tend to blow the returning oil out of the filler-cap.

In cases where no connection is provided for arranging the feed to the by-pass filter, it is necessary to take advantage of other sources, of which there are usually two alternatives. These are, the external pipe to the oil-pressure gauge, and the external oil-feed to overhead valve gear in o h v engines. In either case, it is possible to obtain suitable unions which may be substituted for those already on the pipes, where they connect to the engine. The replacement union will be of T-formation, providing an additional point of attachment for the connection to the filter.

In some cases it may be necessary to restrict the by-pass flow through the filter, so that the pressure throughout the oil supply system is not reduced by too free a passage through the alternative circuit. The filter makers provide a special adaptor with a small-bore hole, for insertion at one end of the pipe connection, the hole being graded to pass an adequate amount of oil to the filter but not enough to affect the pressure.

The connecting pipes

It is always advisable to use flexible piping for the pipes to and from the filter, and obviously this is essential if the filter itself is mounted on, say, the bulkhead, or battery box or toolbox, as the flexible engine mountings will cause relative movement which must be accommodated by the piping. Special tubing is, of course, necessary, of a pattern which will stand the considerable pressure on the feed side of the filter. Mention has already been made of the various ways of leading the oil back to the sump. If it is decided to fit a union to the valve-cover on an o h v engine, take care that the union is put in a position where the flow will not cause too much oil to reach certain parts. For example, it is undesirable for the flow to be directed over one of the valves, as the lubricant may swamp the oil-sealing device on the valve stem, and cause oiled-up plugs. The most suitable position is to lead the oil into the cover at a point where there is an adjacent drain down to the sump. Near the push-rods will usually meet the case, as the push-rod tunnel often forms the return oil passage in this type of engine. The cover-plate used to enclose the valves on side-valve engines (and the push-rods on o h v) forms another useful point into which to lead the return supply, but to remove this cover calls for more dismantling work than is necessary with the rocker cover. (It is perhaps needless to add that in no circumstances must any attempt be made to drill any such covers *in situ*.) When refitting the cover-plate after operations, it is necessary to take extra care over joints, as there will be somewhat more oil about than the original design anticipated, and thus more chance of leakage.

Chapter 7

The Cooling System

The cooling water—Flushing—Leakage—Hose clips—The thermostat

The cooling system of the average car does not call for much attention to keep it in good condition. This is just as well, since in very many cases it receives no attention whatever until such time as trouble starts. In winter, and also on the odd occasions when the temperature is above normal in summer, cooling system faults which have been developing as a result of neglect are likely to assert themselves.

It should be appreciated that the cylinder water jacket, radiator, and associated components, form a very important part of the mechanism. It is not just a case of the cylinders being surrounded by water and thereby being rendered incapable of overheating or other ills. The heat given to the cylinders by virtue of engine operation is not distributed uniformly, and there are odd corners where intense heat is generated; for example, in the vicinity of the exhaust ports. A well-designed system takes care of such items by giving a definite direction to the flow of water. In this way, an ample supply is taken to the hotter places, while other parts, as for instance the lower parts of the cylinder bores, are induced to maintain a constant temperature by a relatively stagnant water flow.

It will be obvious from the above that even if no sign of boiling is apparent in the form of steam from the radiator vent, it is quite possible for local overheating to occur. This can well result in damage to exhaust valves and their seats; even a small amount of distortion will prevent the valve seating absolutely correctly, with a corresponding loss of power. In addition, of course, any form of local overheating is likely to cause pre-ignition.

The cooling water

The main drawback to the use of water for cooling is its liability to deposit a crust on the jacket surfaces, the result of continual heating and cooling. The amount of deposit which accumulates is governed by the characteristics of the water, and thus varies in different parts of the country. It is not unusual for instruction books to recommend that the radiator be always filled up with rain-water. Unfortunately, water-butts are not normally a feature of modern housing, and it is a human failing that the rule is to use the tap-water at hand, for better or worse.

Actually, this is quite satisfactory as a general rule but, where particularly 'hard' water has to be used, extra precautions are called for. In such cases, the system should be flushed out about once a year with one of the advertised radiator-cleaning compounds. This will ensure that the deposits do not build up to excess. If it is made a rule to do this before putting in the anti-freeze in preparation for the winter season, the job will take little extra time, and will not be overlooked.

In connection with the use of anti-freezing mixtures, there appears to be some doubt as to the advisability of using these all the year round. Actually there is nothing against the presence of the mixture as such, in the system. The only point is that periodic topping-up of the radiator with water will obviously reduce the protection given, and if this is continued indefinitely all through the calendar, one may be caught napping when benefit is most needed. The use of fresh anti-freeze at the start of each winter (or the reinsertion of the mixture drained after the previous one) will cut out this risk, particularly as in the latter case, a modicum of the same brand would be used to restore the percentage in the mixture, before reinsertion.

Flushing

When flushing the system it is advisable first of all to remove the drain taps completely from both radiator base and cylinder block (Fig. 7.1). Before flushing, the engine should be allowed to cool if it is at working temperature. When the water has all drained off, fresh water should be run from the hosepipe into the neck of the filler for about fifteen minutes, or until the outflow from the drain tap holes is clean. In very obstinate cases of choking, it is often possible to clear the radiator tubes by removing the filler cap and bottom hose connection, and forcing water from a hosepipe into the bottom stub. An adaptor of the type shown on Fig. 7.2 is used by garages for this purpose, though something less ambitious but equally effective can be readily made up. The idea is that by reversing the flow of water in relation to its normal passage, a great deal of foreign matter will be dislodged. Provision must of course be made to deal with the waste water emerging from the filler neck, to ensure that it does not spill over the engine.

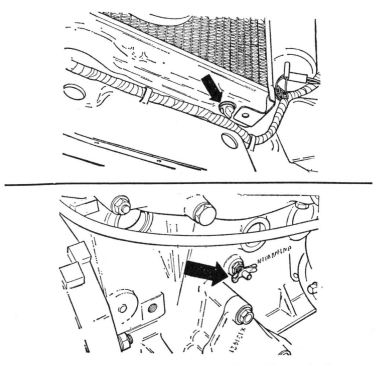

Fig. 7.1. Drain taps or plugs are usually found on radiator and cylinder block, as shown.

The old type of cooling system has the radiator overflow pipe arranged so as to be open to the atmosphere and to take the water straight to waste. It is thus possible for sudden braking to cause a surge of water down the overflow, and frequent topping-up may be necessary. Modern systems have a filler-cap of the type shown in Fig. 7.3 which incorporates a spring-loaded valve, arranged to open at a pressure of about 6 lb per sq in. This maintains the whole of the cooling system under pressure, and not only prevents any untoward loss of water from causes such as that just mentioned, but also raises the boiling point of the water so that the engine may work at a more efficient overall temperature. When the engine is warming up the valve opens due to expansion in the system, and allows air and water to escape. On cooling down, the vacuum valve in the filler-cap opens, and air re-enters the system. If the overflow pipe is led into a container holding about a pint, the cooling system can be regarded as 'sealed' (to use the popular term) with no topping-up required. The pipe is arranged to go down right to the bottom of the flask, so that when water is expelled under expansion,

Fig. 7.2. Hosepipe adaptor for 'reverse-flow' flushing of radiator.

it enters the bottle instead of being wasted. When the system cools down, the resulting vacuum draws the water back into the radiator. Apart from cars which are equipped with this arrangement as standard, the overflow bottle and associated parts are available in kit form, for fitting to popular makes.

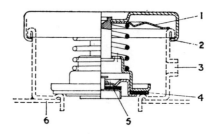

Fig. 7.3. Section through radiator filler cap, as used on most modern cars.
1 Friction plate. 2 Retaining lugs. 3 Pressure release pipe. 4 Pressure valve seal. 5 Vacuum valve seal. 6 Header tank.

Leakage

A leaking cooling system is a menace, and is quite inexcusable. The overflow from the radiator is the only place at which an escape of water is permissible, and unless this is of bad design the overflow should be slight. There is no point in over-filling the radiator but, even if this is done, the first few miles will drop the level to the correct degree. If on inspection this appears to be unduly low down

THE COOLING SYSTEM 135

in the filler neck, do not immediately top up again. Have another look a few days later (so long as there are no leaks) and it will almost invariably be found that the level has remained as before.

This matter of over-filling is of particular importance when anti-freeze is in use as the mixture is by no means inexpensive, and there is no point in throwing it away down the overflow.

Another possible source of leakage is at the water-pump shaft gland. Older cars had a shaft gland of the adjustable type, consisting of the traditional form of stuffing-box, using some such substance as graphited asbestos string for the packing medium. It is, of course, possible to repack the type without much trouble and to obtain the requisite degrees of tightness on the gland by means of the adjusting nut. Nowadays, the type is of interest only to vintage-car owners, who will be well aware of the technique required to obtain a leak-proof packing without imposing too much friction on rotation of the shaft.

The modern water pump gland is of the non-adjustable type, the water seal being effected by a spring-loaded packing ring inside the pump body, as shown on Fig. 7.4. The seal assembly is obtainable as

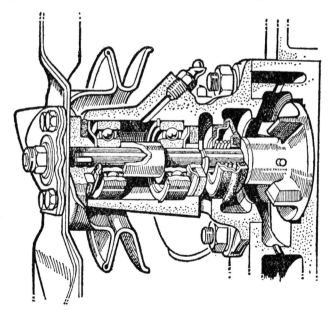

Fig. 7.4. Sectional view of typical water-pump with non-adjustable sealing gland.

a replacement item, but the actual dismantling of the pump though quite simple, invariably requires the use of special tools not likely to be found in the home garage. Thus when a leak starts, the best remedy is to have the pump replaced by a service reconditioned unit on the usual 'exchange' basis. It is most inadvisable to run the pump for any length of time with a leaking gland, as if the seal breaks up altogether, the loss of water will be quite serious.

Some pumps have a completely 'sealed for life' bearing assembly, requiring no extra lubrication, while others have a nipple or plug on the body as shown on Fig. 7.5. If this attention is required it should not be neglected.

Fig. 7.5. External plug for periodic lubrication of pump bearings, on Morris-Oxford engine.

Hose connections are the next point. These are generally forgotten until they begin to leak but are easily checked during routine maintenance. The modern flexibly-mounted engine allows a lot of relative movement between the water take-off pipes and the radiator, and in some cases this movement is absorbed by only a short piece of large-diameter hose. The presence of cracks in the outside of the hose, however slight, is a sign that it should be renewed. When undertaking this task, the correct hoses must always be used, particularly if moulded to a special shape. It is of course possible to bend the modern convoluted hose quite remarkably without kinking, but nevertheless the type should never be used if a properly shaped connection is standard practice. The latter looks more of a job, and passes the flow of water in a more efficient manner.

Hose clips

There should be no difficulty in fitting new hoses if the pipes have been properly cleaned and all traces of the old hose, and shreds of rubber, scraped off. To enable the new hose to be slipped over, it is a good idea to apply a little soap to the inside of the hose, when it will go into position without difficulty.

There is a right way to fit hose clips. They should be positioned on the hose with their edges, say $\frac{1}{8}$ in. and no more, from the end, and tightened until they grip firmly, but no more (Fig. 7.6). There is no

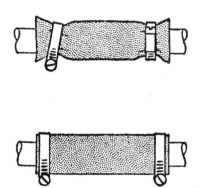

Fig. 7.6. The wrong and right way to fit hose-clips.

point in using such force that the hose assumes a strangled appearance; this only encourages leaks. If the clips on each hose are fitted with their bolts, or other fixings, all in the same plane, they will look right, and when deciding just where to position the bolts in relation to the surrounding parts, give some thought as to the ease or otherwise of tightening and removing them, which any particular position will provide.

The thermostat

The thermostat unit is housed in close proximity to the outflow where the water is at its highest temperature, which is usually at the front of the cylinder head. It comprises a plate valve actuated by a heat-sensitive bellows or wax capsule surrounded by water. When cold, the plate valve is seated, restricting the water flow and enabling the engine block rapidly to attain a working temperature. When the water warms up, the valve gradually lifts off its seat, allowing a free passageway for the water. Fig. 7.7 shows

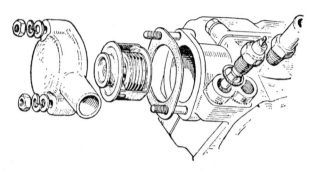

Fig. 7.7. Thermostat and housing at front of cylinder head on Triumph Vitesse engine.

the thermostat removed from the water take-off point on a typical engine. The thermostat is arranged to open at a water temperature of about 80 deg C. and to be completely open at 95 deg C, both figures being approximate. If a fault is suspected, the unit can be tested after removal by immersing it in a water-filled container along with a thermometer. The water is then heated, and the opening temperature noted. If defective, the unit must be replaced, as there is no practicable method of repair or adjustment.

CHAPTER 8

Electrical Matters

The electrical system—General principles—Live wiring—Circuit fuses—Current capacity—No-fuse wiring—Wiring and accessories—Soldering aids—Other connections—Cable runs—Wiring security—Insulation of cables—Earthing

In Chapter 3 the importance of the electrical installation in relation to efficient ignition was stressed. There is another aspect now to be dealt with, which is liable to increase the demands on the generator and battery (and thus the overall power unit load), namely the addition of accessories which, when correctly installed and used, add to the pleasure of motoring and assist safe driving.

Emphasis is frequently put on the ease of installation of such additions, and it is true that, superficially at any rate, there is nothing in fixing, say, an additional spotlight after a fashion so that it will work when switched on. However, like most other things connected with motor cars, there is a right way of doing this sort of job, and it is proposed to suggest a few simple ideas which will enable something better than the 'flex-and-insulation-tape' layout to be achieved.

The electrical system

The standard electrical system is nowadays so reliable that owner-maintenance is virtually confined to keeping up the level of the battery liquid, and ensuring that no corrosion is taking place on its terminals. Beyond this, a periodic check of the generator-belt tension and (possibly) lubrication of its end-bearing, is the total attention called for. Thus, in comparison with earlier days, many owners have only a very sketchy notion of how the system operates. This does not matter too much; the hazard lies in the little that they do know! For example, if it is required to connect up an additional electrical accessory, one has only to find a live terminal to obtain a source of electrical energy. The main battery terminal is an obvious one, but there are several others, notably the 'sw' terminal on the ignition coil, the starter switch, and so on. Take a connection from one of these, through your extra switch, thence to the new gadget, lash the wiring to anything handy with insulation tape, and the job is done.

Wiring done in this manner may last for years; it may more likely give trouble. In the latter case, failure is liable to be sudden and devastating, involving a complete loss of electrical energy throughout the car. Since it is almost equally simple to do the job properly, the risk is not worth while. Incidentally it should be mentioned that makeshift wiring is by no means confined to amateur jobs; many garage electricians excel at it.

General principles

For a start, we can consider some points relating to the source of energy—the battery—and the significance of the connections. All electrically-operated components require two connections to achieve a continuous current flow (the system being of course uni-directional, or DC). In the case of the car, one connection is formed by the metalwork of the vehicle which is termed an 'earth'. Both positive-earth and negative-earth systems are in vogue and it is important not to get the connections wrong when dealing with units embodying transistors and diodes, such as radios, clocks, impulse tachometers and alternators. Reversing the polarity can quickly burn their circuits out.

The battery positive terminal, marked with a cross and sometimes coloured red, is the one from which current flows out of the battery when the latter is feeding the electrical system. It is also the terminal by which current flows into the battery from the charging generator. The battery negative terminal, marked with a line or dash and sometimes coloured black, is the one by which, in the two cases mentioned above, current returns from the electrical system to the battery, or returns to the generator from the battery.

It should be made clear at this point, that the general DC system of the car is unaffected if instead of a DC generator being used to supply the charging current, an alternator or AC generator is substituted. The latter has certain advantages (which need not be considered here), particularly when high electrical output is called for. Alternators are provided with accessories which convert the alternating current to DC at an early stage, and thus the charging supply to the battery remains unaltered.

From whichever battery terminal is used, the earth connection is made of very stout braided copper strip or similar material, and is taken to a point on the engine or car structure where there is plenty of metal. It might be thought that there is no complication about making an earth connection, but as a matter of fact it can be a source of considerable trouble. Major faults in arranging the main battery earth would be to connect it solely to an engine mounted on rubber blocks, but this has been done. A little thought will show that, to get from the engine block to the car in this case, the current will quite possibly have no path other than those formed by the propeller-shaft and axle, the carburetter controls, exhaust pipe brackets, and so on. In each case, the current has to traverse a path of doubtful electrical conducting virtue; grease-filled bearings and joints, dirty flexible brackets, etc.

Live wiring

The non-earthed or 'live' battery terminal is connected to the main insulated cabling throughout the car. From it, a very stout cable runs to the starter switch. The switch terminal which receives the cable is sometimes made the 'take-off' point for the other insulated leads to the panel on the instrument board; from this panel go the various lighting and auxiliary circuits (direction indicators, wipers, etc.). The charging circuit, comprising the generator, voltage regulator, and cut-out, is really a separate layout and is unlikely to require attention, as it is not affected by any additional electrical 'gadgetry'.

At one time the electrical system contained a good many fuses; the various circuits fed by the switch panel were each fused separately, so that a short-circuit in, for example, a headlamp cable, simply blew that particular fuse without affecting anything else. This made for some ease in tracing the faulty circuit, though the job was difficult enough on a dark night! Modern wiring, however, very rarely gives trouble, and nowadays it is usual to confine the protection by fuses to the auxiliary circuits; the lamps are wired direct to the switch panel, without fuses.

Circuit fuses

Further to this elimination of lamp fuses, the number is often cut down to two, both of these being arranged to protect the small accessories containing fine-gauge coils, such as the wiper motor, and which in the event of a fault developing might smoulder for a long time without any sign of trouble, with the eventual risk of a 'real' fire. The fuse will of course prevent this. However, it is usual to arrange one fuse to protect several such accessories, so that in the event of trouble with, say, the wiper motor blowing the fuse, one might find the direction indicators also out of action.

The small number of fuses fitted as standard nowadays are also connected in a special manner which has some bearing on the way in which extra electrical gear is connected. One fuse will be wired through the ignition switch, as a concession to owner-forgetfulness in the way of leaving things 'turned on'. It is usual to wire several of the more frequently used accessories via the ignition switch, so that whenever the engine is switched off, these items will also be rendered inoeprative, irrespective of whether they are separately switched off or not. Typical of such liable-to-be-forgotten fitments are the wipers, and heater fan motor. The coil and electrical fuel pump connections, however, are usually independent of this fuse.

A second fuse will be arranged to protect other items while still allowing them to be used whether the ignition is switched on or not; these might include the horn, cigar-lighter, etc. Adjacent to each fuse is fitted a terminal, to which the extra accessories may be connected.

Current capacity

The two fuses may be marked as 'Aux' or 'Aux Ign' as shown on the control box illustrated on Fig. 8.1. This is a pattern which is

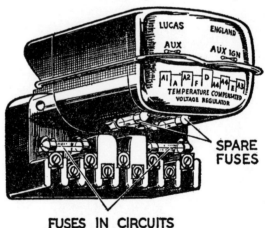

Fig. 8.1. Control box of earlier type with two marked fuses and two spares.

becoming less used, and many cars have a separate fuse-block of the type shown on Fig. 8.2. If required, additional fuse-blocks similar to this can be fitted to cope with extra electrical fittings. It will be appreciated that apart from any extras to be added, the existing fuses may already be coping with a fairly heavy current; thus it is usual to find that they are fairly liberally rated, that is, they may have a carrying capacity of 30 amp or more. They still have adequate protective value, of course, as any trouble almost invariably means a fairly sharp and heavy battery discharge which will effectively blow the fuse in question. When all the accessories on one fuse

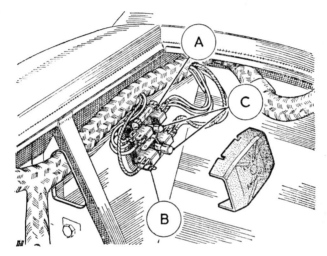

Fig. 8.2. Separate fuse-block with fuses and spares, as used on Mini-Minor.
AB=35 amp fuses, C=spare fuses.

are in action simultaneously, there may not be as much margin for 'extras' as might be thought, and a rough check can be useful. Dealing with accessories wired through the ignition switch, the heater fan may take 5 amp, the wipers another 4. Thus, on a cold and rainy day, we have a steady flow of 9 amp. Every time the direction indicators are operated, another 3 amp goes, making a temporary increase to 12 or so. Even thus, there is plenty of margin so long as things are not overdone. On cars not provided with heaters, a likely additional accessory would be an electrical de-mister, and this is an obvious example for wiring through the ignition switch. The device is usually provided with its own built-in switch, but this is easily forgotten when leaving the car. A de-mister can take a current of

5 amp or more and, if left on for a full day, will make a lot of difference to the state of discharge of the battery.

No-fuse wiring

Extra-loud horns and powerful spotlights are quite often wired through fuses and, if the latter are already loaded with standard accessories, little margin may be left. In fact, however, neither lights nor horns, if correctly wired, have any real need of fuse protection, and there is a strong case for leaving the fuses to look after fittings that really need them. A further point worth mentioning is that, as the fuses are usually wired through the ammeter (if fitted), a heavy-current horn connected thereto will give rather a nasty 'kick' to the ammeter needle every time it is operated; all to absolutely no purpose, but prejudicial to the well-being of the ammeter.

For such accessories independently of fuse protection, a connection is taken straight to the 'live' battery terminal. This sounds quite simple, and is, but the job must be done correctly, as we are now tapping the reservoir, as it were, direct, with no protection in case of accident. (Not that wiring through the fuses should be made any less carefully.) Before going on to describe the principles of good wiring there is one accessory often fitted to older cars that should have special mention; that is, the reversing light made out of an old headlamp or similar fitting. If inadvertently left 'on' at night, this can be one of the most irritating and dangerous of gadgets but as it is usually connected via a simple switch on the instrument panel, depending on the driver's memory to operate correctly, it is little wonder that one is occasionally compelled to trail behind a forgetful user for miles.

When no operating switch is fitted to the gearbox reverse selector, the job is obviously not easy to make foolproof. The only solution, as British law requires, is to have an additional tell-tale light in the panel near the switch, and of adequate brightness, wired in parallel with the reversing light.

Wiring and accessories

A complete range of aids to correct wiring is now available at accessory shops, and a really professional job can be made by anyone who will take the trouble. For virtually any extra fitting, however simple, it is probable that most of the following will be required. First, a length of insulated cable. This must be of the correct automobile pattern; no household wiring, whether flex or 'permanent' type, is in any way suitable. The correct cable is fairly heavily insulated, but its main feature is the tough outer sheath, which resists damp and chafing against securing clips, etc. It is quite flexible, but should not be bent carelessly as it may be possible in this way to crack the sheath. Then, for making the end-connections to bolt-type terminals, small 'spade-eyes' are necessary. These have a hole, of a suitable size to go over the terminal screw, typical sizes being 4 BA, 2 BA and ¼ inch clearance holes. The terminal shank has claws which are clinched with pliers over the bare portion of the cable end,

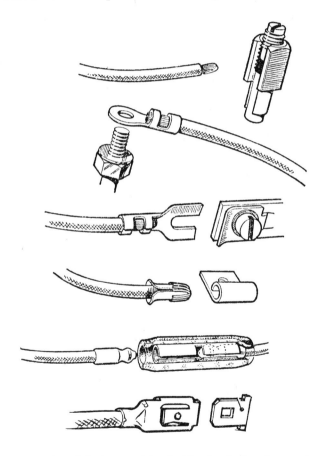

Fig. 8.3. A few of the large variety of wiring terminals and connectors now available.

the final joint being made by solder run liberally over the whole shank.

Plastic or rubber insulating sleeving is useful when two or more cables have to be taken over, under or around parts of the car where they cannot be firmly secured but with which they may occasionally make contact. However, when laying out the wiring run, it is essential to choose a path, whenever possible, that will allow the cable to be clipped rigidly throughout its length.

A selection of terminal fittings is shown on Fig. 8.3 and it will be noted that in some cases, the connection is made by clinching or squeezing the metal body over the strands of the cable. This practice is widely adopted in the motor trade, and is satisfactory if expertly done with the correct tools. Nevertheless, it is considered that a first-class job really warrants the use of solder, on the grounds of absolute permanence and an extra margin of safety. It is obvious that two wires can be joined simply by twisting their bared ends together, and when everything is clean and bright the full quota of current will be passed. Slight corrosion, however, can cause a surprising amount of extra resistance; the sure safeguard is to 'tin' the twisted joint with solder, so firmly uniting the two wires and giving them a corrosion-proof covering, which is then insulated with tape.

Soldering aids

The job of soldering has been made very simple nowadays by the availability of solder-sticks of 'wire' having a flux core; thus, solder and flux are applied simultaneously to the work, and the old trouble of only having two hands no longer applies. The only basic and important requirement is that the parts (terminal tag and cable, or whatever applies) shall be clean. Small parts for electrical connecting are usually 'tinned' as supplied, while of course the copper wire strands take solder very easily. As regards the iron, a self-heating iron of electric type is ideal; the amount of heat necessary is however quite small, so that the older kind of externally-heated soldering 'bit', if of reasonable size, will readily cope with several small electrical jobs at one heating, assuming a reasonable amount of agility on the part of the operator.

Apart from joining two wires, or attaching spade-eye terminal tags, which have already been mentioned, other items which it is desirable to solder are, in particular, the small tubular cable-ends which are often used on horns and spotlights. These are arranged to be push-fit into a spring clip on the accessory body, and the 'lazy' way of connecting is to spread the wire strands down the outside of the tubular portion and jam them between it and the clip. In this way, only a proportion of the strands carry any current, while a minor tug will part the connection, apart altogether from the corrosion risk already mentioned. The proper way is to pack the end of the stranded cable into the tube, doubling the wire back on itself if necessary to fill the tube. Then heat up the whole assembly and run solder liberally inside the tube. Such a connection will be there 'for keeps', and has the further advantage that the tube can be slid into its clip on the horn, etc., or unclipped, without prejudicing the joint in the slightest.

Other connections

There may be odd accessories in which no terminals are provided, and it is necessary to solder the cable directly on to a metal tag. In such cases, as much length of wire as possible should be allowed in contact with the tag, and the strands should be spread so that the solder can get hold properly. Care must be taken not to apply too much heat to the tag, if it is apparent that adjacent parts, such as coils or insulation, may be affected.

After fitting any kind of attachment to a cable end it is advisable to reinforce the insulation at this point; this will add some strength, as well as extra insulation; both are important, as it is often impossible to avoid a fairly sharp bend just at the point of connection, for example in the cramped quarters of a lamp body. A short length of insulating sleeving of the plastic variety, or a turn of black insulation tape neatly applied, is all that will be necessary.

When two cables have to be joined by twisting and soldering, black tape is a perfectly satisfactory wrapping, but there is no need to go to extremes in building up a 'cocoon' of this, which makes a very untidy job. One or at the most two thicknesses of tape, which overlaps the cable insulation for half an inch on either side of the bare joint, is quite adequate, with one proviso; that the cable so treated is firmly clipped in position, and not liable to chafe.

Cable runs

The way in which a cable is laid out on the vehicle at once shows up the 'professional' job. It is quite easy to take a wire from terminal 'A' to accessory 'B' by the quickest and shortest route, so that it intrudes at uncalled-for places and looks what it is, a makeshift addition. By using more wire and planning the route carefully, it is usually possible to run the connection in such a way that it is hardly seen at all. Where possible, a workmanlike method is to follow the path of the existing electrical connections. The latter will probably be encased in a braided 'loom' containing several circuits, and of course there is no question of interfering with this. However, supposing such a loom is taken to the front lamps, or some of them, an extra cable for a spotlight in the vicinity could probably be taken along the outside of the loom to the front of the car, and need only be 'on its own' for a very short distance.

The loom itself would no doubt be secured by adequate clips to the bodywork, and a similarly neat method is called for in attaching the new cable to its outside. For this purpose, lashings of tape are definitely barred. A variety of clips is available, useful types being the light-alloy strap and buckle, which is passed around both loom and new cable and fastened in exactly the way one expects from its description. A similar fixing is obtainable made of rubber, but on the whole the metal ones look tidier. Incidentally, the single cable can often be tucked away under the loom and out of sight—another mark for neatness.

If it has been possible to follow an existing cable run for a portion of the way, it will obviously be necessary at some point to break off for the final connections. Here again, there must be no appreciable length of unsupported cable left on its own, metal clips being used for security to within a short distance of the final terminals.

In cases when it is necessary to run the cable where there is no other existing wiring to act as a 'master-guide', it is more than ever necessary to plan carefully. First of all, it can be taken for granted that if the wiring is correctly carried out, it will give no trouble; thus, there need be no hesitation in putting it out of sight, as this makes for neatness. For example, if an extra switch is mounted at instrument-panel level, it is usually possible to take the cable from this, closely along the back of the dashboard or scuttle-plate, and quite possibly behind the 'trim' in the footwell. Thus no wiring need be visible in the compartment at all. Where the cable passes through from the interior to the engine-space, it is usually possible to find a rubber-bushed hole capable of accommodating it. If it is necessary to make one specially, this must of course be properly bushed with the correct type of rubber fitting known as a 'grommet'.

Wiring security

The reference to using an existing rubber-bushed aperture should not be taken too literally, as it is undesirable to mix electrical connections with such things as oil-pipes, carburetter controls and steering columns, all of which may well pass through this kind of bush. Non-metallic items such as windscreen-washer pipes or heater connections may provide the answer if sufficient room is available where they pass through the bulkhead, but care must be taken not to have the cable trapped, and to make quite sure that any grommet fitted really protects the insulation from sharp metal edges. If there is any doubt, an extra insulation sleeve, a few inches in length, should be slipped over the cable at this point.

A further tip to note is that it is by far preferable to make a hole specially for the cable, rather than add a lot of length (and untidiness) just for the sake of using an existing through-way. As regards securing the cable, a workmanlike way is to use metal clips, securing them to the panelling by self-tapping screws of small size. The clips are of the type which go right round the wire, and can be opened out to pass over it; a sort of pipe-clip in miniature, in fact. Some people may be reluctant to drill the panels specially for such a relatively minor matter. If there are any existing studs, or bolts with nuts, in the near vicinity, which can also be used to accommodate the cable clips, this is all to the good, and it is worth varying the cable run slightly to take advantage of them. Otherwise, self-tapping screws in specially drilled holes are infinitely to be preferred to lashing the wiring to odd parts of the engine and frame with insulation tape.

Insulation of cables

When metal clips are used in this manner, a turn of insulation tape should be put around the wire where it is gripped by the clip; this not only augments the insulation but makes for a secure location. It is also necessary to make sure that the clip has no sharp edges, which may bite into the cable. Where two or more cables are being run to the same point (such as a pair to the front of the car, for fog and spotlights, or a group to the facia, where a new switch panel is being located), black insulation tape should be used to bind the cables together, this being wound on puttee-wise, as neatly as possible. It is usual to do this after all the connections have at least been 'trial-fitted', but if it is desired to bind up a set of wires before fitting to the car, the use of differently-coloured insulation enables the ends of an individual wire to be identified, this being of course in accordance with 'professional' practice. A variety of colours is available for cables of otherwise identical electrical characteristics, obtainable at accessory shops.

When accessories are purchased new, the cable entry is invariably well looked after, as regards insulation and watertightness, by some form of rubber device, grommet or 'boot', and beyond fitting the cable carefully and correctly, no further attention is called for. If 'used' equipment is being installed, such devices may be missing or unserviceable. The main thing to take care of is that the cable enters with no danger of chafing the insulation against a sharp edge. If necessary, extra insulation by means of rubber sleeving or tape may be added, the aim being to ensure that the cable is a good tight fit. Inside the accessory (for example, a spotlight), sufficient free cable should be left to enable the connections to the lampholder to be made comfortably, without straining the cable or making it liable to pull off as soon as the lampholder is fitted into the casing. Where the cable emerges from the accessory, it must be supported as close up to the latter as possible, so that there is no chance of vibration causing the cable to creep and put a strain on the terminal connections. This is a far better method than tying a knot in the cable inside the lamp!

Earthing

As a final word, the question of earthing is most important. We have all seen, or experienced, the lamp which only lights when kicked or thumped, and some motorists are apparently quite happy to accept this phenomenon. The cause is of course a poor earth connection to the car, usually the result of rust. The remedy is to rely on something better than the fixing screws of the accessory, to provide an earth. Some electrical items nowadays have an earth terminal for 'optional' use; if such is available, take a cable from it to the nearest sizeable bolt or nut on the car that can be unscrewed, and clamp the end of the cable, via a spade terminal, under this. If no earth terminal is in evidence, clamp the cable under one of the fixing screws of the accessory.

Do not earth to bumper bars, or other items which have a lot of bolted joints between them and the car, as each joint represents a potential loss of current. Obviously the nearer the earth-connection can be taken to the battery, the better.

Chapter 9

Seasonal Welfare

Climatic conditions—Other handicaps—About the battery—Methodical charging—Generator and wiring—The starter motor—Bearings—Sticking pinion—Ignition ills—Extra insulation—High-tension circuit—Water ingress—The carburetter—Fuel pipes—'Taxi' operation—Radiator connections—The thermostat—Filtration and cooling—The ignition system—Vapour locks

After summer, a percentage of everyday car users begin to lose any reputation they may have had for punctuality in their comings and goings. Instead of their place in the regular procession to and from town, individuals are to be seen engaged in unaccustomed tuning operations in their driveways, or even endeavouring to push-start. Immobilised vehicles with clouds of steam issuing therefrom are occasionally seen at the roadside. And there is the really unlucky one who arrives in a taxi or falls back on the charity of his fellows. But is he unlucky or merely neglectful?

Climatic conditions

Most, if not all, of the foregoing misfortunes are due to the climate; it really is as simple as that. What puzzles many drivers is the ill-effects which even a few degrees drop in temperature can have; they could understand it better if troubles started with the snow, that is, when it is really cold. However, facts have to be faced, and although it is easy to reproach the unfortunates with lack of proper maintenance, it must be agreed that during the summer and early autumn, when all is well, there seems little point in doing seemingly unnecessary jobs and perhaps running into extra expense. Having thus acknowledged the frailty of human nature, we can now consider just what causes these 'autumn blues'.

It is safe to say that the biggest culprit is the battery. This is no reflection on a component that is extremely hard-worked at the best of times, and frequently neglected. It is not always appreciated that the best of batteries, by the very nature of its operating sequence of continual charge and discharge, is expendable. The load of modern electrically-operated accessories means a heavy drain, which must be balanced by a correspondingly heavy rate of charging. The starter today is, quite legitimately, expected to start the engine without any preliminary hand-cranking such as was virtually insisted on (with a warning of trouble otherwise), by manufacturers many years ago. Thus, the acceptance, with good grace, of a battery life of about two years is justified; it is most decidedly not a device on the part of battery makers to increase their trade.

Other handicaps

As, when attempting to start, the battery supplies current both for rotating the engine and for the ignition, its influence is considerable. Thus, any minor defects in the ignition system which would cause little trouble under favourable climatic conditions, become more serious if the applied voltage is decreased due to an extra current drain on the battery. Then again, cold, damp air is no help to mixture formation for starting; small inaccuracies in carburetter settings, in respect of the starting-mixture and slow-running adjustment, exert disproportionate influence under these circumstances. And finally the quality of both fuel and lubricating oil can make a lot of difference; the former in regard to its facility for vaporising under 'compression' heat (the only heat which is available initially) in a cold cylinder, and the latter because freedom from 'stiction' enables the engine to rotate that little bit faster.

Now, consider one of the vehicles belonging to an unfortunate driver such as we mentioned at the commencement. It has run satisfactorily throughout the best part of the year. Its battery is getting on for two years old. Its plugs have done something over 20,000 miles —they should have really been thrown away a few weeks ago. The ignition system generally was last checked over several months ago, and the various components, plug leads, coil, etc., having never had so much as a wipe with a duster, are covered in a film of oily filth. On odd occasions during the summer, the slow-running has been erratic, and the nearest available screw on the carburetter has been twiddled to effect a cure. Thus, the general carburetter adjustment is suspect. To end on a more charitable note, we will assume that at least the engine is mechanically sound, and that good quality fuel and oil is used.

Now, with the foregoing, all or in part, it only requires a very small drop in temperature, which is almost invariably combined with a moist atmosphere as well, to cause starting trouble, ranging from a minor disinclination to get going to complete immobility.

About the battery

We said that the battery is usually the main culprit. We can now investigate what can be done to ensure freedom from anxiety in this department.

First there is this question of battery life. Short of waiting until that cold morning when the battery simply jibs at turning the engine, or gives up the ghost after a few turns, how can one tell just how much life is left? The stock answer usually given is that, in the event of a slowing-down of battery power such as this, the battery should be charged at a garage, and tested with various instruments. The writer's experience is that such procedure generally results in a few days' return to vigour, and then another lapse. He would also suggest that there is no instrument in existence that can indicate how good a battery is, *i e*, in terms of its ability to go on performing satisfactorily.

After all, the first query should be—why is the battery suddenly jibbing at doing what is its normal task? Assuming that it has been kept properly topped up, properly charged, and the terminals and connections clean, the only reason for a sudden lack of power must be internal trouble. The only satisfactory remedy for this is a new battery; it is possible to find repairers who are prepared to rebuild a faulty cell, and in the case of a twelve-volt battery with six cells, this may appear an economical method of repair. The snag is that one 'dud' cell usually means that the others are about to go the same way. The usual form of internal trouble is an accumulation of sludgy matter which comes off the battery plates; this eventually collecting in sufficient depth at the bottom of the container to short-circuit the plates. As well as this, however, the shedding of active material by the plates themselves reduces their charge capacity. This effect is, of course, minimised by the construction of first-class modern batteries.

In regard to correct charging, it is a fact that short journeys combined with considerable use of lights plus increased starter loads may tip the balance of charge against discharge to an extent which results in a gradual discharging of the battery. This can obviously produce symptoms not unlike those of a 'dud' battery, in that, if a low state of charge coincides with a particularly cold morning, the starter may have that sluggish manner suggesting lack of energy. Although the high rate of charge of a modern generator will restore the debit very quickly in normal conditions, it will be appreciated that quite a lot of current is taken by some of the more-used accessories (apart altogether from headlights and spotlamps). If, each day, only a little more is taken out than is put back, the time will inevitably come when, only half-charged, the battery will fail to do its job.

Methodical charging

It is not at all difficult to tell whether a battery is fully charged or not, and this without waiting for a minor calamity such as failure to start. If all is well, for instance, the brilliance of the headlamps, with the engine stationary, should be very little below that which obtains when the engine is running and the generator charging. If the beams 'perk up' as soon as the engine is revved, it shows a low state of battery charge and, if this persists for any length of time, it indicates that there is no surplus current available, in the particular conditions of operation, to liven up the battery.

The remedy for this is simple: install a trickle-charger in the garage. This item is worth its weight in gold to most drivers, but particularly so to those who must have unfailing reliability day after day, and have no time to be bothered with 'service' batteries and the nuisance of having to fiddle about at the local garage while batteries are changed over. A good trickle-charger is not at all expensive, and consumes little current from the house mains. A suitable two-pin socket is connected permanently to the battery, and mounted either on the facia panel or under the bonnet; thus, it is only a matter of a minute to plug in the charger as soon as the car is put away in the evening.

The popular type of charger for home use has an output of about 1 amp. This sounds very little, and obviously in terms of current consumption it is. (A rough calculation shows that for a consumption of 1 unit, a 12-volt battery could be charged for about 30 hours.) Nevertheless, in the time between say 7 p m and 8 a m, 13 amp-hours will have gone into the battery, and this is more than sufficient to make up the deficiency caused by any normal excess of output in relation to input. Occasionally, too, there is much to be said for giving the battery a really good 'refresher' charge at a low rate; this can easily be done, when a charger is available to hand, during an odd week-end when the car is not used.

Finally, the charger can be a very present help in cases of forgetfulness; lots of drivers have occasionally left some electrical accessory on all night, perhaps a defroster, heater, or similar accessory, which is quite capable of completely running down the battery before morning. A charger plugged into a flat battery and left for half an hour or so will enable the engine to be started on the handle, particularly if the charger is left in circuit while getting a start. There is a legend frequently heard, that a low-output trickle-charger is incapable of charging a battery from scratch, *i e*, one that has been completely discharged due to a mishap such as the foregoing. There is no foundation for this story. Obviously, however, when a battery is in such bad shape that it leaks current away faster than the charger can put in, such a situation might arise to lend substance to the theory.

Generator and wiring

The generator and its regulating device are so reliable that owner-attention should normally be confined to lubrication of the bearings, if this is called for by the maker's instructions (Fig. 9.1). More

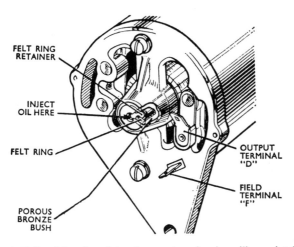

Fig. 9.1. Tail-end bearing plate of generator, showing oiling point for porous bronze bush bearing.

expert attention, such as cleaning of the commutator and replacement of brushes, is of course well within the scope of the knowledgeable enthusiast. At one time this attention was necessary at fairly frequent intervals, in order to keep the generator in a state of full efficiency, and the technique was thus well worth acquiring. Nowadays, however, the generator will maintain its full output without such attention for 25,000 miles or more. If all seems well after a mileage such as this, it is a good idea to take a look at the brushes, just to see that they have not worn to a point where they are beginning to disappear down into the brush-boxes or holders. At the same time, while the cover is off the machine, any carbon dust present can be blown out; with the modern ventilated type of machine, very little of this accumulates.

The driving belt should be kept nicely tensioned, but not so tight that it will not slip when the fan blades are pulled by hand. Too tight a belt overloads the bearings of both generator and water-pump (in cases where both are driven by the same belt) and is more liable to break. Too slack a belt can often be detected by fluctuations of the ammeter needle, and in this condition the belt will get very hot due to slipping. A new belt may stretch a little in the first couple of thousand miles, but after being adjusted after this mileage, there should be no necessity for any further attention for a very long time.

Never attempt to take up the adjustment of the belt at regular intervals as a sort of maintenance routine, but only when it is quite obvious that attention is required.

The generator wiring should need no attention beyond making sure that the terminals are tight. These are sometimes rather inaccessible and slight slackening-off can lead to peculiar effects such as an intermittent refusal to charge which can conjure up visions of all kinds of trouble. Once properly tightened, and with suitable lock-washers fitted, there will be no such worries. As regards the battery terminals, some attention is necessary. The heavy cable-ends which locate on the battery posts must be quite clean initially, over the whole of their contact area, the same applying of course to the posts. After fitting, a smear of petroleum jelly or one of the special compounds sold for the purpose will prevent corrosion.

With the battery and its charging arrangements in first-class order the only other item concerned with rotating the engine is the starter motor itself, plus the switching arrangement. The amount of current taken is very large and the connecting cable is therefore of heavy section, and kept as short as possible to cut down resistance. Thus, it is not usual except on old cars to mount the main starter switch on the dash panel, as this would call for long cables. The switch proper is either mounted on the starter motor carcase and operated from the dash by a mechanical pull-control, or else it is of the solenoid-operated type, located in a convenient position and engaged by a separate electrical circuit through a suitable switch on the facia.

The starter motor

Like the generator, the starter motor calls for little routine attention; in any case it is usually so inaccessible that owner-maintenance is often impracticable. As it handles very heavy current, its electrical construction is exceptionally robust, and when trouble does occur it is usually very obvious, and calls for complete overhaul. The best way of ensuring a trouble-free starter motor is to see that it gets its proper quota of current. Insufficient electrical power means that the starter 'jibs' or almost stalls, at each compression stroke, as it turns the engine over. At these points, arcing then takes place between the brushes and commutator. This produces 'flats' and builds up an electrically-resistant deposit on the commutator, so that the passage of current is further hampered and the effect is progressively worsened.

The only remedy for this trouble when it occurs is to have the armature removed and overhauled, together with new brushes. In normal use, the latter wear extremely slowly, and very rarely require renewing. However, it is a useful precaution, occasionally, to remove the cover-band from the starter at the commutator end, check the brushes in much the same way as was advocated for the generator, and blow out any stray copper dust with a tyre-pump.

Bearings

The armature-shaft bearing at the commutator end is almost invariably of the plain bush type with self-lubricating properties, calling for no attention. It is occasionally found that the bush sets up a high-pitched squeal when the starter is in action, generally due to partial drying-up of the lubricant with which the bearing is impregnated. Though unpleasant this noise is not harmful. It can be cured by a small application of penetrating oil to the end of the shaft, this working its way into the bearing without difficulty. The remedy should not be overdone, as it is essential that no oil is allowed near the brushgear. At the driving end, the quick-thread 'inertia' mechanism which engages the starter pinion with the flywheel occasionally gives trouble; this again can often be laid at the door of a defective battery or switch. The engagement depends essentially on an instantaneous 'kick' being given to the armature shaft, so as to start the pinion moving along the thread, and into mesh with the flywheel. Obviously, the best chance of obtaining such an impulse is under the influence of a full quota of current. However, the driver can help by ensuring always that movement of the starter switch is quick and decisive (thinking about something else while operating the control first thing in a morning accounts for a surprising number of stuck starters!). Also, once the starter is operating, keep it in action until the engine is firing with reasonable urge—don't release at the first feeble explosion.

Sticking pinion

At one time it was necessary to give periodic attention to the pinion engagement mechanism, or 'Bendix drive' as it is popularly called. This was usually because even a small amount of dust or greasy matter on the thread of the shaft was sufficient either to prevent the pinion running-up into flywheel engagement, or once engaged, to prevent it coming free when the engine fired. Washing in paraffin at intervals could usually be relied on to keep things happy, and this is of course still the remedy, if the components do become fouled in this manner.

However, modern clutches throw off very little dust from their linings, and the clutch mechanism calls for little or no lubrication; added to this, the design of the drive has been vastly improved over the years, together with a more powerful initial kick of the starter. Thus, trouble is unlikely from dirt or oil. When sticking occurs nowadays it is generally due to a combination of circumstances which are really nobody's fault, and which need not be gone into, so long as it is understood that trouble in this context should rarely happen.

When this occurs, the old remedy, which cannot be beaten, is to engage top gear, and rock the car vigorously backwards and forwards. The effect of jerking the engine flywheel in opposite rotation to normal (which is what this action does) will free the pinion, almost invariably. An alternative is to turn the armature shaft by means of a spanner or wrench on the squared extension of the shaft which is provided for this purpose (Fig. 9.2). The extension is sometimes protected by a detachable cap, which must first be removed. The normal position of a starter motor does not make this job of turning by hand very easy; however, it may be considered less undignified than the first suggestion!

As the final aid to positive starter engagement we come to the switch. When operated by a pull-control, it is of the lever type, with simple flat contacts pressed firmly against each other by a toggle movement. So long as there is no fouling by oil or dirt, no attention is called for. In the case of switches fitted directly on the starter body, it will be appreciated that dirt is liable to accumulate, and attention to cleaning is well repaid. As far as solenoid-operated switches are concerned, there is little for the owner to do beyond ensuring that the terminals are clean and tight. The electrical mechanism of this type of switch is definitely not one for owner-attention, and in the event of persistent trouble, a service replacement unit should be fitted. It is worth mentioning that if the solenoid circuit fails (*i e*, the circuit to the dash control) the starter can be energised by pushing the solenoid plunger by hand; it is usually located under a rubber cover at one end. Another method is to short-circuit the

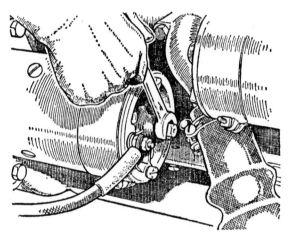

Fig. 9.2. A jammed starter can sometimes be freed by turning the squared shaft-end with a spanner.

two heavy terminals on the solenoid switch with a piece of metal. As sparks are likely, however, this should only be done where there are no fuel leaks, and by reasonably expert hands.

The foregoing attention as well as having ensured that the engine rotates at a reasonable starting speed, will also provide an adequate margin of ignition current. Assuming that the ignition system is maintained as detailed in Chapter 3, there should be little to cause 'seasonal' trouble. However, even at the risk of a little repetition, some possibilities should be considered.

Ignition ills

The main cause of partial or even complete failure of the ignition is moisture causing current leakage. It is amazing how very little damp can render an engine completely immobile, but many readers must have experienced the seeming miracle, whereby an engine that has resolutely refused to give any sign whatever of starting has fired straight away after the high-tension leads have been wiped over with a dry cloth. It is worth any owner's while to gaze carefully at his ignition system, and consider the possible number of leakage paths between terminals, along the outside of the insulation; he will be surprised how many there are.

The first necessity is cleanliness of the insulation. A film of dry dust, though undesirable, does little harm during the summer, as engine heat keeps the moisture content down; with winter rain, fog, and damp generally, however, such a film becomes well impregnated, and in fact forms an excellent electrical conductor. It will be appreciated that the film need only extend from a bare terminal to the nearest metalwork (for example, from one of the coil terminals to the body of the coil) to form a leakage path. The first requirement therefore is simple; keep all surfaces of ignition parts such as the top of the coil, distributor cover, spark plug insulators, and all leads, quite clean.

It will be obvious that even with this precaution, a film of moisture is still quite capable of forming, $i\,e$, on a perfectly clean surface. Though this is less likely happening, it can occur, but of course evaporation is rapid once the engine warms up. The problem in this case is therefore to ensure that the minimum leakage occurs through this 'clean' moisture.

Extra insulation

The terminals on ignition apparatus of current types are shrouded to a greater extent than used to be the case. This shrouding (by moulding the insulating material around the terminal, for example), extends the leakage paths, and also tends to put a break in it, so that odd spots of moisture cannot combine into a continuous circuit. It may be, however, that due to the position of components, exposed terminals on the low-tension side are particularly vulnerable. Obviously, leakage from this circuit, operating at battery voltage, is in no way comparable in its effect to leakage of high-tension current, but it is still worth preventing. If considered desirable, therefore, low-tension terminals, such as the two on the coil and the one on the distributor side, may receive additional protection. This can take the form, as a rather crude measure, of plastering some kind of putty-like compound over the whole terminal.

Such compounds are now available for this specific purpose, but the material is used in exactly the same way as was the Plasticene beloved of old-time reliability-trialists, the waterproofing qualities of this famous substance being a byword. Alternatively, moulded rubber protectors can be fitted at the ends of the leads to combine with the terminal-tops, this of course making an effective and neat assembly.

High-tension circuit

The high-tension circuit comprises the leads from coil to distributor, and from the latter to the plugs. Although all these leads usually appear to be well served with protective terminals, there may be room for improvement.

Starting at the spark plugs, these are nowadays almost invariably provided with some form of snap-on connector at the lead-end, the body of this being of insulating material. Such connectors could be completely protective, were it not for the fact that some examples use an exposed brass set-screw, put in from the outside, to secure the h t lead into the body of the terminal. Obviously, this constitutes a possible contact point for leakage. As the screw is usually recessed well into the body once it is home, the remedy is to fill up the recess above the screw-head with shellac or similar material.

Some of these terminals, particularly on old side-valve engines 'built to a price', give a false sense of protection; the plug position on such engines is particularly vulnerable, as any stray water tends to find its way to the cylinder head. Difficult starting, and mysterious misfiring in a rainstorm, can often be remedied by the substitution of a first-class type of plug terminal, which completely shrouds the plug insulators, in place of the rather skimpy 'umbrellas' sometimes used. These plug protector-terminals should not be left in place indefinitely. Of course, if examination of plugs is carried out as a routine measure, removal of the terminal and cleaning of the plug insulator is automatic. However, plugs are often left *in situ* for astronomical mileages and, in such cases, it is advisable occasionally to pull off the leads and clean from the insulators the accumulation of oily dust that collects there.

At the distributor-end of the leads, the point to watch is where the leads enter the moulded distributor body. It is usually a very close fit, one secured either by an insulated gland-nut or by an internal set-screw, or even a spring clip. It is possible for moisture to find its way between the cable and the entry, in which case it may not only build up a leakage path, but will also corrode the contact in the body. This latter may be the cause of a misfire that is difficult to trace. However, moisture can be excluded from this vulnerable point by running shellac or similar compound between the cable and its hole. Obviously this will make the cable difficult to remove but, in any case, such a measure is only called for when the cable needs renewing, so the job is worth doing.

For extreme cases where it seems impossible to keep the leads free from a moisture film (due perhaps to a combination of 'no garage' and climatic environment) it is possible to obtain a water-proofing dope which, when applied to the whole of the exterior insulation, renders it much less susceptible to the wet. In the ordinary way, however, this remedy should not be required, and must not be regarded in any way as a substitute for proper insulation on the lines already detailed.

Water ingress

Ingress of water, during heavy storms, to the inside of the engine cover is an annoying fault to which some cars are prone. It is difficult, to some extent, to design a bonnet-cover which while providing adequate engine ventilation at the same time prevents an excess of water from entering. Nevertheless, most manufacturers manage to do it, but there are odd vehicles, often of early design, on which this is a real problem. In these cases, it is sometimes necessary for the unfortunate owner to lift the cover and drape a waterproof sheet over the engine block, before leaving the car for the day in a car-park,. Apart from the 'primitive' aspect, and the nuisance of this precaution, there is the very real danger that the sheet may be forgotten in the evening hurry, and could easily catch fire from the heat of the exhaust manifold, or get tangled up in the fan.

It is fairly certain that the judicious use of thin sheet aluminium fitted inside the bonnet sides to form deflector plates could be arranged to prevent any water reaching the engine, without prejudicing ventilation. If, for example, the leak is at a central bonnet hinge, a U-shaped aluminium trough under the hinge could be arranged to collect leakage, and to lead it harmlessly to the rear of the bonnet where it would drain away.

Apart from its effect on the ignition system it is of course possible for too much water to affect carburation also, should it collect in sufficient quantity to gain access to the float chamber (via air-vents). However, we will assume that the necessary steps have been taken to prevent all this, and now go on to the final requirement for winter reliability—correct carburetter adjustment.

The carburetter

In the warm weather peculiar things sometimes happen to carburetters; the owner may decide that because of the prevailing temperature, he can do something about economising fuel. He then carries out some adjustment towards this end or asks his garage to do so. This sort of thing may or may not affect the starting and idling adjustment, depending on the make of carburetter, but the chances are that it will. The first essential, therefore, in making the carburetter do its share towards easy cold-weather starting, is to have the starting and idling settings made absolutely correct.

The procedure is fully described in Chapter 4. Assuming that the settings are corrected as necessary, there are other matters which should have attention. It is essential that a full head of petrol is available for a quick start; when the fuel is supplied by an electric pump, this is ensured by waiting until the pump has filled the carburetter, before engaging the starter. Also, in this case, impending pump failure is usually heralded by excessively noisy operation caused by air leakage through a punctured diaphragm.

When a mechanical fuel pump is used, driven from the engine camshaft, several engine revolutions are 'used up' in filling the carburetter. This should not impose an excessive drain on the battery, so long as the pump is in good order, as the latter is of quite high pumping capacity. If, however, a mechanical pump is on its last legs, with a defective diaphragm and possibly leaking valves, it may call for far too many engine revolutions before the carburetter is filled, and all this wasted effort is obviously unfair to the battery. It is easy to check, by using the hand-priming lever, whether this type of pump is in serviceable condition; as both electric and mechanical pumps are available on an exchange basis, at a moment's notice, there is no excuse for continuing with a 'dud' one.

Fuel pipes

Obviously, petrol pipes must be free from leaks, and all unions tight. Otherwise, apart from any other objections, excessive pumping work is called for. Also, of course, the whole of the fuel system must be clean, and if the filters have not been looked at for a long time, this should be done now.

Finally, there is the important matter of controls. The starting control at the driver's end may concern the main jet, as in the S U carburetter, a separate starting carburetter (Solex) or a strangler flap. In all cases it is essential that the full control movement at the driver's end is faithfully transmitted to the carburetter. Excessive backlash in flexible cables may accumulate over a period, which prevents this. It is then impossible to gauge to a nicety the degree of control operation, apart from the fact that in some cases lack of full movement may actually prevent the mechanism from functioning.

So much for the descent into winter. To continue on perhaps a happier note in regard to climatic conditions, we will consider what attention is necessary when the evenings become lighter and the air gets a little warmer. First, it will be safe to suggest that at this time there are plenty of cars in operation which have the anti-freeze in the cooling systems, and which in other ways have not quite shaken off the lethargy of strictly utilitarian winter motoring. As the days lengthen and holiday plans are made, however, there is every reason for a spell of maintenance attention which will ensure trouble-free operation over the long mileages and in the higher temperatures which can be expected during the next few months.

Providing the car was reasonably well prepared for the winter, there should be no need to ferret about looking for signs of rust-ravages; the exception is that if the vehicle has been operated in an area subject to much snow, it will be as well to make an extra underneath examination. Modern 'underseating' compounds are extremely effective but some of the solutions put on the roads to assist in melting the snow are also extremely efficient at removing paint, etc.

'Taxi' operation

If the last few months' motoring has been confined to what might be called 'taxi' usage, this implies fairly low speeds, frequent starts and stops, and little drastic call on the brakes, or hard working of mechanical parts. In such circumstances it is a fact that the engine (and vehicle generally) settles down into a rather 'lazy' condition, machinery being curiously human in these respects! Thus the tonic effect of a few simple jobs is well worth having, from the point of view of performance alone, apart altogether from the reliability aspect.

We commenced by mentioning anti-freeze, and the treatment might well start from here. It is becoming more usual for people to leave anti-freeze in the system all the year round, on the basis that it does no harm and saves money. The point has already been emphasised in Chapter 7, that frequent topping-up with water must progressively dilute the solution. So long as the radiator and water jackets are reasonably clean and free from deposits, however, this is the only real objection to the practice. Many cooling systems require little topping up, and those of 'sealed' type with an overflow bottle, none at all. Thus the question of anti-freeze protection comes down to maintaining the cleanliness of the cooling system and the strength of the solution, and the latter again depends on the type of system. Garages are nowadays equipped with a sampling device whereby the strength of the solution can be readily gauged, and this is the answer. If the solution has become unduly diluted by topping-up with water, it is easy to bring it up to strength by adding a little anti-freeze.

Radiator connections

The hose connections must be carefully examined, and this includes any auxiliary hoses, such as those connecting to the heater. 'Pinching' between finger and thumb will usually reveal whether there are signs of internal perishing of the rubber, and any doubtful ones must be replaced. When anti-freeze is put in, it is not unusual to find a slight leak or two at the hoses, due to the facility with which the solution finds its way through microscopic passages (being much freer in this respect than 'straight' water). The quick remedy invariably adopted is to tighten the clip heartily, but this does not always do the hose any good, and thus, if there is an appearance of strangulation due to this, replace the hose. After filling up with clean water, run the car for about a hundred miles and drain off again; if the liquid is clear, well and good; the system is probably commendably free from scale. If, however, there are signs of rust or other foreign matter, refill, and add one of the good quality radiator cleaning mixtures, strictly in accordance with the maker's instructions. This last remark is made deliberately, as some of these cleaners are fairly searching, and it may be inadvisable to leave the mixture in for more than a limited period.

It should be remembered that short journeys in a low temperature make little demand on the cooling system, and a clogged radiator core or indifferent circulation will have small effect; in fact, they may be an asset in helping the engine to warm up quickly. On long, fast drives in summer, especially in hilly country, efficient cooling is obviously essential.

The thermostat

The thermostat unit is usually very reliable, and unless inspection is required as described in Chapter 7, no action should be necessary. In winter, it is possible for an alternative thermostat to be fitted, which is designed so that the valve opens at a higher temperature, so facilitating rapid warming-up in the colder climatic conditions. If this has been done, the normal setting should be reverted to in summer.

It may be that the attention just outlined will have the unfortunate result of producing the odd radiator leak. Why, it is difficult to say, but such happenings are obviously a triumph for the 'leave-well-alone' school, though the latter are liable to overlook the probability of the radiator completely disintegrating out in the wilds as a result of their recommendation. The remedy for such minor leakages is to use a good brand of radiator cement which is introduced into the water. Here again only the best quality should be adopted, and it will be found to make a permanent repair.

It should be possible subsequently to judge the effectiveness of the cooling by actual driving, but obviously the only real confirmation is provided by a radiator thermometer. If the car is not so equipped, there is every reason for adding this instrument. Boiling nowadays is a luxury that cannot be indulged in with the same carefree abandon as used to be the case when the jacket held several gallons, and prior warning enables the necessary action to be taken. In winter, too, fuel is saved by keeping up to a sensible temperature reading.

The only other likely place for leakage is the pump gland; nowadays this is non-adjustable, and the remedy is a replacement pump, obtainable as a 'service' spare. Any leakage at this point should be dealt with thus, as a slightly leaking gland may break up suddenly, and allow serious water loss.

With the cooling system dealt with as described, the next item calling for attention will probably be the sump contents. Winter driving, with short journeys and indifferent warming-up, does little to improve the engine lubricant; this is liable to become quite appreciably diluted by condensed fuel, while water may also be formed, leading to more sludge than usual. The normal oil-changing routine will be adequate to restore full efficiency; it should be emphasised that if a long holiday journey is imminent, but the oil-change mileage is still not quite 'on the clock', an early change should be made for once, rather than leaving the job to some probably inconvenient time half-way through the holiday.

It is quite usual nowadays to use the same grading of oil all the year round, but if a higher SAE number is specified for summer operation, this should, of course, be adhered to. In elderly engines it may even be advisable in very hot weather to try the next higher SAE rating than that recommended, as oil consumption may well be lessened thereby. It should be pointed out, however, that 'thicker' oil is no substitute for bearing metal, and the only excuse for this practice is to save the overhaul cost in an otherwise good car that is soon to be disposed of for some reason. When using thicker oil than standard in this manner, extra care should be taken not to work the engine too hard during the warming-up period, but to allow the oil to get properly circulating first.

Filtration and cooling

No doubt the oil filter will have been attended to at routine intervals, but here again if a change is about due, the job should be done in good time. As dissipation from the external surfaces of the oil circulatory system is one of the methods by which the engine rids itself of unwanted heat, the outside of the filter and sump exterior should be cleaned. The latter in particular is likely to accumulate remarkably tenacious filth in winter, but a good, strong hosepipe jet will shift it quite readily.

A few thousand miles covered during the winter may well lead to an increase in oil consumption which is not noticeable on short journeys but will show up in long, fast driving. This increase, though quite legitimate, may impress itself suddenly by the necessity to top up the sump after say a couple of hundred miles, to the extent of a pint or so. As this represents 1,600 miles per gallon, it is decidedly not excessive, and should be accepted as normal. However, a useful and reassuring addition to the equipment is a spare quart tin of oil (of the same brand as in the sump) mounted in a bracket for easy access. This will guard against emergency and the anxiety of running with a low sump level; garages are not always open, and even in the British Isles one can still find places where they are few and far between.

The ignition system

The plugs, probably, have suffered somewhat from the sort of motoring already called 'lazy'; a good fast hundred miles will sometimes restore them, but it is preferable to have them properly cleaned. Of course, if they are due for renewal this should be done; 10,000 to 15,000 miles is the normal useful life of the average, inexpensive, non-detachable plug in a popular type of engine; after this mileage, the insulation deteriorates; it is thus not safe to judge the condition by external appearance or by the degree of burning on the electrodes. The more expensive types of plug are in a different category; platinum-pointed types in particular have a very long life, and it is not unusual for them to last about four times as long as the normal type without deterioration. Apart from a good clean externally, there should not be any further attention necessary to the ignition system, except that called for by the amount of running done, *i e*, there is no need to inspect the distributor as a special measure, but if it is due for examination this should of course be done, and the contact-breaker points adjusted at the same time.

Vapour locks

Carburation, again, should be quite in order providing everything has functioned well during the past months. If by any chance a slightly richer setting has been used in the cold weather, it may be possible to weaken the mixture, this applying particularly to the S U type, where a difference of one odd 'flat' on the jet hexagon may be beneficial. Fuel filters these days seem to accumulate remarkably little foreign matter, and the only real source of possible trouble may be in the fuel pump itself, the remedy for which has already been detailed in Chapter 5.

A good deal of nonsense has been written about the effect of hot weather combined with mountain-climbing, in producing 'vapour locks' in the fuel feed, due to evaporation of petrol in the pipes. Such things do occur, usually on Alpine gradients in sun temperatures which are exceptional. In almost all other circumstances, the trouble is caused by a pump diaphragm that is on the point of failing, plus slight leaks at the pipe unions on the suction side of the pump. A contributory factor could be wrongful replacement of a fuel pipe too near a source of heat such as the exhaust manifold.

Chapter 10

Accent on Economy

First, the driver—Engine efficiency—Exhaust loss—Limitations—Carburation and distribution—Carburetter matters—Variety of devices—Automatic Air valves—Good idling—Opinion—Economy jets—Restricted breathing—Air filters

Most drivers when considering fuel economy, rightly assess this in relation to adequate engine power. In some circumstances, however, there may be a call for the usage of as little fuel as possible, even at the disadvantage of a power loss. Although, happily, petrol rationing now seems to be a thing of the past, its ever-increasing cost (due largely to successive governments' fiscal extortions) acts as an effective deterrent to squandering fuel. It is felt, therefore, that a chapter with the accent on economy only will be useful to those who for various reasons are obliged to remain mobile on the minimum amount of fuel.

First, the driver

It is not proposed here to instruct the driver on how to conduct his vehicle in order to obtain the maximum mileage per gallon. It is well enough known by now that the greatest factor in ensuring a too-rapid flow of the precious commodity is what is sometimes referred to as the 'leaden right boot'. As low a speed as can be stomached, with the minimum throttle opening, and the least possible alteration of accelerator position, are the best aids to economy; they are in fact as much as a great many drivers can hope to achieve, it being somewhat futile to suggest acceleration to 40 mile/h and a long 'coast' down to 20 in neutral, if one's daily jaunt consists of two six-mile journeys through a housing estate. Furthermore, while coasting in neutral may cost nothing in fuel, it is doubtful whether, to attain this end, we should throw overboard the safety aspect; travelling out of gear, except momentarily, is rightly frowned on in all official instructions dealing with road safety, and by the expert driver.

However, stunt driving apart, slower speeds, with throttle control as described and a sharpening-up of the powers of anticipation, will make much less use of the brakes feasible. After this, proper maintenance of the vehicle in regard to lubrication and brake adjustment can do quite a lot. Much of this is, of course, left to garages nowadays, but a spot check of the odd maintenance item might be well worth the trouble. The author's gearbox was recently found, after such periodic attention, to contain 25 per cent too much lubricant; a few errors like that can have quite an effect on oil-drag, and thus increase running friction.

At lower speeds, tyre pressures can usually be increased without affecting passenger comfort, but should never be pumped so hard as to hammer the suspension unfairly. The articulating points and bushes in the latter are designed for a specific type of loading, and the rapid, high-frequency shocks caused by hard tyres, though of small magnitude on reasonable roads, can cause rapid suspension wear.

So much for generalities; we can now consider the subject which is really within the province of these pages, that is, whether it is possible to tune the average engine to use its fuel more economically, and, if a really drastic reduction in power is acceptable, whether a proportionately bigger saving can be achieved.

Engine efficiency

In order to start with a clear idea of what is involved, it is necessary to appreciate what happens to the petrol in the tank in its conversion to power at the engine flywheel. Statements are frequently made, informing that most of the fuel is wasted and, naturally enough, deploring the fact. These statements are, taken literally, true, in that only a small proportion of the heat energy in the fuel eventually becomes power. On the other hand, it must be conceded that the conversion of a can-full of liquid into the rotation of a shaft which propels a motor-car is quite a creditable feat in itself! If we exploded the can of fuel in the back-garden, we could say that we had converted liquid into power with remarkable efficiency, but power in that form would be of little use to anyone.

It is not difficult to represent the horsepower at the engine flywheel in terms of heat value. Thus, if we know the amount of fuel being fed to the carburetter, and its heat value (or calorific value), it is a simple matter of calculation to determine how much of this is converted to useful power. The answer in a good engine is around 27 per cent; sometimes more, often less.

Thus the inventor of any device which, to quote a recent advertisement, 'ensures that every particle of fuel is burnt usefully' would have no need to fear for his future prosperity. Unfortunately, however, this loss of the available power in over 70 per cent of the fuel is inherent in the engine operation; the necessity for keeping the temperature sufficiently low to enable it to function without becoming so hot that it seizes means that a considerable amount of the heat value of the fuel is dissipated in the cooling system, whether by continually maintaining a large volume of water at a high temperature, or by radiation from air-cooling finning. In the former case, it can be said that we manage to recoup some of this loss by using the hot water to make the car interior more comfortable. Ideally, of course, the temperature within the cylinder would vary between wide limits according to which particular stroke of the operating cycle was being performed; this is the answer to those who query why a 'Heat engine' should require cooling. The temperature allowed by the cooling system is a compromise, and if regulated in a reasonable way, to ensure that too much heat is not extracted, its usage of fuel heat-value is kept to the inescapable minimum.

Exhaust loss

The next big source of heat loss is down the exhaust pipe. The engine dimensions dictate a piston stroke of a certain length, and it is obvious that if we develop the engine to a state of efficiency whereby the pressure at the commencement of the power stroke is adequately high, there will still be quite a lot of pressure in the cylinder when the exhaust valve opens. If we wish to get rid of the burnt charge in time to allow the engine to scavenge properly (which is essential to high revolutions), a considerable exhaust pressure must be accepted.

The remaining loss, again unavoidable, is that required to rotate the engine and to drive its auxiliaries, and is absorbed in friction. This can best be illustrated by imagining the engine set to 'idle' at 800 rev/min in the normal way. If, with no other alteration, we remove the fan belt, cutting out the fan, water pump and generator loads, we shall find the engine speed becomes appreciably higher, say 1,100 rev/min. If then we alter the carburetter adjustment to bring the speed back again to 800 rev/min, we shall reduce the amount of fuel used. The same rule would of course apply in regard to any other reduction of friction or load in the moving parts of the power unit.

In the above brief description of heat losses, we have deliberately avoided too much complication, and have simply tried to show that, whatever may be inferred to the contrary by purveyors of economy aids, the petrol engine is inherently inefficient, in the sense that most of the fuel is employed in bringing about conditions that enable the engine to rotate, and only the relatively small part that is left produces power! This is, however, no reflection on engine designers.

What we should aim to do, having accepted these losses as rational, is to consider how they can be reduced to the minimum, and it is not difficult to classify the 'desirable' as: production of the most burnable mixture from the petrol–air combination; its delivery to the cylinders in as equal proportions as possible; ignition and combustion to give the most effective power strokes, followed by a clean scavenge of exhaust gases; and finally, reduction of friction to the minimum.

Limitations

In normal circumstances, the average owner is not exclusively interested in mileage per gallon, so long as it is no worse than the next man's. Performance is what he really appreciates; but of course if this can be combined with a good mileage per tankful, all the better. An engine designed for efficient working will perforce use its fuel to the best advantage, and it is rare today to find any engine which is inherently extravagant. The latter characteristic went out finally with the demise of the side-valve type, particularly those of very large cylinder capacity.

In considering the manufacturer's attitude to fuel usage, it should be borne in mind that, apart altogether from the engine and transmission, a public demand for some peculiar frontal shape or bigger tyre section may undo all the good that much redesigning of the induction manifold has achieved, from an economy standpoint. When two drivers in identical cars can return mileage-per-gallon figures that vary by 20 per cent (due entirely to their driving and maintenance methods), it is understandable that where popular cars are concerned, precise attention to thermal efficiency and reduction of friction may take a back seat in relation to economy of manufacture and fool-proof control systems.

Despite this, however, such vehicles are in fact remarkably efficient, as shown by their test-bed performance.

Attachments to the carburetter and modifications to the induction system have attracted the attention of those who seek after greater economy, for a long time; this is quite logical, in view of the fact that any existing system falls far short of perfection. Many of the faults, however, are as near as no matter unavoidable, while others only appear as defects to one particular group of user—and the car has to be built to cater for all.

This last remark can readily be amplified. The normal popular-car engine incorporates a hot-spot device enabling rapid driving away from cold, and quick warming-up. Once warm, however, the excessively heated charge mitigates against induction efficiency, because it takes up increased volume in the inlet passages without any extra weight of oxygen. However, the driver who starts up many times a day, and whose engine rarely attains a normal temperature for long, will welcome the hot-spot, while the motorist who warms up once and then drives a good many miles non-stop, will wish for a more economical induction system, even at the expense of some easy-starting virtues. If the engine was produced with a non-hot-spotted inlet manifold, the foregoing positions would be reversed; the frequent starter would curse the time taken to coax his engine into smooth running ease, and would incidentally use quite a lot of fuel in fiddling with the choke; while the long-distance man would welcome the power-with-economy touch afforded by his 'cold' manifold, even though he also suffered some inconvenience in warming up. In an effort to obtain the best of both worlds, some engines

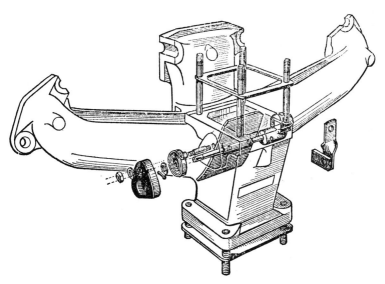

Fig. 10.1. Exhaust manifold with hot-spot incorporating thermostatically-controlled flap valve.

have hot-spots of the type shown on Fig. 10.1; these are put out of action by a thermostatically-controlled device when the engine has attained its normal temperature. It is unlikely, however, that such designs can ever be as efficient as a system devoid of hot-spot, in terms of power for fuel used, as manifold shape is bound to suffer at the dictates of hot-spot incorporation, even when the latter is automatically rendered inoperative.

Carburation and distribution

As far as the carburetter itself is concerned, there is usually little amiss with its functioning. Its size is chosen by the engine designer with regard to the volume of mixture that it is required to deliver, in order to enable the required power output to be obtained. The component parts, jets, choke tube, and so on, are graded to the same end. It is most important that this is fully appreciated, since a good deal of nonsense is talked and written about the possibilities of obtaining greater economy of fuel by altering jet sizes, admitting more air, and so on. If greater economy is achieved by such methods (and this is quite possible), it will invariably be at the expense of power production, which is of course a natural result.

As has already been stated, it is virtually impossible to obtain equal filling of the cylinders from one carburetter. In this respect, the more carburetters there are, the better, but inequality will still be present, the nearest approach to the ideal being, of course, the use of one choke per inlet port and no manifold, as on racing cars. Not only does the shape of the manifold prevent equal delivery of mixture, but minor differences between the individual cylinders will affect the degree of negative pressure available to draw in the charge. In arriving at the optimum mixture strength, ignition, timing, compression ratio, and so on, the most efficiently charged cylinder will determine the settings, while those not so well-filled will be producing less power than would be possible if all could be equally matched.

Carburetter matters

Coming back to the carburetter for the moment (since it is after all where mixture production commences), we can consider whether it is in fact possible to effect a saving in fuel without loss of power. This implies that the carburetter, already of the correct size and properly tuned to the engine's requirements, is left severely alone.

We now have to look for shortcomings which it may be possible to alleviate. The principal one concerns the bias given to the mixture flow by the normal butterfly-type throttle valve, which tends to throw the mixture to one side of the carburetter outlet, allowing some collection of liquid fuel on the side of the passage. A good hotspot will of course offset this to some extent by 'frying' the liquid, but the result can well be a rather patchy mixture.

The interposition of a gauze screen or similar device between the carburetter and the manifold will obviously assist in keeping wet fuel off the wall of the duct; in fact, liquid will tend to cling to the screen, where it is picked up by the air-stream. Such devices are usually arranged to be fitted between the carburetter-to-manifold joint. The form of the centre-portion may vary from a piece of ordinary copper gauze to honeycomb meshes, and even miniature propellers and vanes designed to agitate the mixture on its way past. Such fittings have the common disadvantage that the through-way of the carburetter bore is restricted, and this must mean a loss in volumetric efficiency. However, occasions when an engine is driven for any length of time with the throttle fully open are rare indeed. One device overcomes this objection to some extent, by substituting for the screeen (or like component) an artificial 'bore' of slightly reduced diameter, carrying an annular knife-edged lip on its top or 'entering' edge. This forms in effect a second choke tube at the manifold entry, the lip collecting stray fuel drops which are then carried away in the main stream of mixture by a graduated air-bleed passing the back of the bore. Apart from the small reduction in through-way diameter caused by the inserted choke-bore, there is no added restriction to mixture flow.

For maximum power of course it is desirable that a clear throughway should exist from the carburetter choke-tube onwards, when the throttle is wide open. Thus, a device which, whatever its virtues in other directions, blocks up the bore to some extent, will inevitably cause some power loss, assuming that the whole induction system has been designed primarily with maximum volumetric efficiency in mind. There are a few such manifolds, notably on the best class of sports car engines, but as we have shown, the normal set-up has to satisfy many other requirements (including, not infrequently, those of the foundry making the manifold) and it is quite on the cards in these cases that the resultant better quality mixture will compensate for any minor blockage of the through-way. In any case, as has been pointed out, vehicles of this type do not operate very frequently on full throttle.

There is a further hazard arising from the location of additional hardware in the inlet passageway. In carrying out its intended function of atomising the droplets of fuel, they tend to cling to the screen or whatever and, if weather conditions are suitable, icing up will occur.

Variety of devices

In assessing the merits of the several types of device for fitting in the position under discussion, it is a good idea to follow two rules: first to consider the reputation of the maker and the length of time that the device has been on the market; secondly (having, it is hoped, by now obtained a working idea of what is feasible and what is not, in the way of improvement), to judge carefully the claims made for the particular fitting, with regard to its specification. The more extravagant such claims are, the more suspicion should be attached.

Reliable sources indicate that on an engine in good condition, an improvement of 5 per cent or even more, in the mileage covered per gallon, can be obtained using one or other of these fittings of reputable make, and this is very often accompanied by smoother running.

The extra-air inlet which admits additional air into the manifold at a point beyond the carburetter is another very old-established economy device, which was quite popular in the days before carburetters had reached their present satisfactory state of design, and when drivers had more time to give attention to such of them as were operated by hand. The modern versions are usually designed, like the fittings already dealt with, for insertion between the carburetter-to-manifold flange. They operate automatically once the amount of air entering has been set by hand to meet particular requirements of the carburetter.

While a small amount of extra air probably does no harm, it should be borne in mind that this will enter only in conditions of very small throttle opening, when there is considerable depression in the manifold; air will not bother to flow through a minute additional orifice when there is a wide-open carburetter bore to go through. If a small air entry is beneficial at 'idling and just over' speeds, this rather points to something amiss with the idling and intermediate settings of the carburetter.

There is nothing similar nowadays to the type of extra-air valve marketed (notably by Bowden Wire Ltd.) a great many years ago, when the driver was able to give some attention to operating such a device. The degree of skill required to operate to the best advantage is against its use; apart from that, the modern engine is not of a type on which the use of haphazard 'extra air' should be encouraged. The result of unduly weakening the mixture in an attempt to obtain economy could well result in damaged valves or other ills, resulting from too much heat or prolonged burning in the cylinders as a consequence of the slow combustion of a weak mixture.

Weakening in this manner is of course very different from providing the carburetter with jets which give an economy setting. In the latter case the mixture is still under full control of the throttle, engine power being restricted accordingly. A separate hand-operated air-valve, however, can be opened up with the engine running on full bore, and the sudden transition from a correct to an abnormally weak mixture may have serious results. Thus, anyone contemplating making up his own air-valve with hand operation should have a very clear idea of what is involved in its use in various conditions of engine operation. In other words, it is a luxury for the expert only; with this proviso, and under open-road conditions, a degree of fuel-saving might be achieved, this depending on the extent to which the additional mixture-strength range caters for conditions not already looked after by the carburetter's own compensating system.

Automatic air valves

There are, or have been, several extra-air valves marketed, all designed for automatic operation, either under manifold vacuum influence or in some other manner; one device is electrically operated. All appear to have one common characteristic; the amount of air admitted is extremely small in comparison with that admitted by the old hand-operated devices when fully open. With this restriction of the air admission orifice, it is evident that little air will pass on full throttle, so that the large volume of mixture will not be diluted very much. On partial throttle openings however, with manifold depression at a high figure, the maximum flow through the valve will take place; since in such conditions the carburetter will be delivering only a small quantity of mixture, the percentage of air dilution will be high.

Obviously an extremely weak mixture will be produced in this way, but as the total volume available to the engine is small, little power will in any case be forthcoming, so that any extra heat generated is easily disposed of, and no damage need be feared. The effect on running must, however, be considered; if the engine idles and accelerates cleanly from the shut throttle position initially, it is logical to assume that drastic weakening of the mixture in this range will upset things. It may do so, but on the other hand cases are not unknown where manufacturers have in desperation drilled a small hole or two in the induction pipe (out of sight of the critical student of design) in an attempt to obtain a good idling and pick-up, even on engines equipped with first-class carburetters.

It is interesting to observe that extra-air devices are beginning to find favour again, but with a different motive from that of economy —the control of exhaust emissions. Further information is to be found in Chapter 12.

Good idling

If the engine is already a 'good idler' it is a sign that there is little wrong in any department. Faults not only in carburation and distribution, but also electrical and mechanical ones, are liable to show signs which can be interpreted by the discerning driver, when the engine is idling or accelerating from very low speeds. If, therefore, any carburetter adjustment is required after fitting an extra-air device it can be taken as a bad sign. Where some form of adjustment is provided on the device, the instructions will probably be to the effect that attention to this will obtain the required results without any alteration to the carburetter settings. Careful investigation will then quite conceivably show that little or no air is passing through the extra-air hole.

Those who wish to try one of these automatic devices would be well advised, before finally making a choice, to ask the maker a few perfectly legitimate questions; these should be readily answerable if the fitting has been properly designed, and tested on a running engine against a brake. For example—performance of the engine at various loads, and fuel consumption in terms of horse-power-hours, before and after fitting. Quantity of air in cubic centimetres passing through the extra-air orifice at various degrees of throttle opening and engine load.

To some, such questions might appear as rather unsporting; well and good, if the purchase is regarded as a sort of gamble. If, however, some of the advertised statements are to be taken at their face value they should be capable of substantiation without trouble, since the data leading up to the final result of 'twenty per cent more m p g' must presumably be on record somewhere.

Opinion

It is a good principle for the technician to try anything, however unpromising it may appear in theory; engineering history is full of such triumphs over scepticism. If, however, the ordinary motorist is expected to experiment with extra air, at his own expense, this is a serious reflection on our carburetter manufacturers, who have had considerably over half a century to perfect their wares. The writer does not believe that the latter have fallen down on their job; in other words, a correctly maintained and adjusted carburetter will produce the right mixture without external aids. As to how many carburetters are in this condition, that is a different matter; but it is no fault of the makers.

Having covered the subject of extra-air admission to the manifold independently of the carburetter, we can now consider what it is possible to do to the latter without adding such unorthodox fittings.

Economy Jets

If a definite power loss is acceptable (and in certain circumstances most drivers are prepared to do this), it is possible with most modern carburetters to change components so that, in effect, the carburetter delivers less mixture generally, and further, that the mixture delivered tends towards weakness instead of richness. In certain phases of carburation, it should be mentioned, the mixture is deliberately enriched in order to provide the desired acceleration, the balance of fuel and air flow being graduated in the appropriate manner.

In tuning a carburetter for economy, it is useless merely to change, say, one part such as a main jet, in the hope that top speed only will be affected, with all else remaining as before. The modern instrument, though trouble-free, is quite complex in its use of metering orifices for both fuel and air; further, the size of the through-way or choke may well have to be reduced if other orifice sizes are revised for less flow. The only way of obtaining satisfactory performance of a reduced order is in effect, to recalibrate the carburetter. In practical terms this means consulting the carburetter-maker's service station, obtaining the new components, and fitting the lot, in place of those already installed.

It should be appreciated that the physical size of the carburetter has been selected for the normal power requirements of the engine, and there is a limit to the degree of economising that can be carried out by this substitution of components. Probably the mere fact that the engine is delivering less power, and that this effect can be felt when operating the throttle pedal, serves as a constant reminder to the driver to employ his usual 'economy' driving methods; at all events, quite good results follow, almost invariably, from the use of these modified carburetter settings, if they are carried out properly. Too much must not be expected, however, and anything in the region of a saving of two or three per cent can be considered good. In this connection, too, the mistake should not be made of adding together all the possible percentage fuel savings which might result from various modifications. This might well come to an astronomical figure but unfortunately it just does not happen! It is nevertheless true that several minor economies can frequently be combined to good effect.

Restricted breathing

It is frequently possible to fit a smaller carburetter, either by the use of a tapered adaptor, which merges the smaller bore with the existing manifold bore, or by fitting a complete new manifold system, which itself may be designed on more efficient lines than the standard one. In this way, the amount of mixture available to the engine can be reduced to almost any desired figure. The cost of such a modification is apt to be rather high, unless good second-hand parts are available, or the owner has access to a machine-shop and is a proficient designer, in which case the adaptor is not normally too difficult to make up.

Air filters

No fuel saving will be effected by removing the air filter, providing it is kept clean; the restriction to air-flow of a modern filter is negligible. It is necessary to ensure that installation is correct, and that connecting hoses, if any, are not allowed to become perished or kinked. Also, of course, the intake to the filter itself must be completely clear, and must not be in too close proximity to heater ducts, bulkhead panels, and so on.

It is probable that as far as sheer fuel saving is concerned, an entry of warm air is preferable to cold. Warm air tends to help carburation, particularly under conditions of light-load operation where the engine as a whole is probably running cooler than usual. Also, the warmer the air the sooner the engine will pull away from cold, and this is more important in the circumstances than having the maximum weight of oxygen contained in the volume.

It is important not to overdo the engine temperature. Some blanking-off of the radiator, particularly in cold weather, is certainly beneficial, but it is unwise to run near boiling point which it must be remembered is considerably above 100 deg C in modern pressurised cooling systems. The reason is that if a high temperature is recorded on the gauge, it is quite on the cards that certain areas such as in the vicinity of the exhaust valves may actually become overheated due to formation of steam pockets; the results might be serious. Hit-and-miss assessments of engine temperature are bad practice, and a thermometer should be fitted if any rational attempt at temperature control is being made—and this is worth doing.

This chapter has emphasised economy at the expense of performance. There is, however, no harm in repeating what was stated in the introduction, that an engine in 100 per cent condition responds to precision adjustment and tuning with benefit both to speed *and* economy. This is the first step, and nothing written here is intended to override that fundamental fact. But if special circumstances make a power loss acceptable, the information will be useful from that aspect, and no other.

CHAPTER 11

A Chapter on Gadgets

Boosted ignition—Low-tension matters—Resetting of plugs—Spark-gap terminals—Distributor details—Extra air—Mixture agitators—Intake silencers—Special intakes—Alternative filters—Carburetter changes—Exhaust pipe fittings

So far in this book we have dealt with items having a considerable effect on the correct functioning of the power unit which, generally speaking, can be dealt with by the methodical driver who is possessed of neither comprehensive workshop equipment nor extensive technical knowledge. Thus, the programme up to now has covered the usual maintenance tasks, but has endeavoured to emphasise the desirability of carrying them out in a workmanlike manner.

It will now be useful to review the more important aids to power which are marketed from time to time, and which year after year attract the attention of even the most sober family motorists at the Earls Court Show.

The advantage of these accessories, and in fact one of their main attractions from the non-mechanical driver's point of view, is that relatively little trouble is usually involved in fitting, and in many cases the claimed advantages seem to imply a very good return for such effort (mental, physical and financial), as may be called for.

In considering such items it is necessary to point out that the final assessment of their worth is the effect on an engine which is already in efficient tune. An engine which is below par can hardly fail to benefit from, say, a change in the make of spark plug, if the substituted set is new. Equal benefit would be obtained, all other things being equal, by cleaning the existing set, the change in make having nothing to do with the matter at all. However, assuming that all is well, the advantages or otherwise of special fittings can be considered.

Boosted ignition

We can usefully start with the ignition system. There are on the market several ignition coils known variously as 'sports' or high voltage types. They are designed to supply their output at about 25 to 33 per cent higher voltage than the standard article, thus enabling wider spark plug gaps to be used if desired, and providing a hotter spark even across the normal gap.

It will be of interest at this juncture to comment, in the light of practical experience, on statements which have been made from time to time, to the effect that so long as the spark will jump the plug gap (at its correct setting) and the engine starts and performs satisfactorily throughout its rev/min range, there can be no advantage in so-called 'hotter' sparks. The author's experience simply concerns an engine which did all these things, but on which, purely as an experiment, the magneto was changed for another instrument which was known to be of an improved pattern. It should be emphasised that in effecting the change, nothing was disturbed, in fact the plugs were not even removed from the cylinders, and there was no alteration of timing, as the shaft coupling was identical on the two magnetos.

The result of the change was that a hill previously requiring third gear was easily surmounted in top. Obviously there could be only one explanation for this—more effective firing of the mixture, even with the same plug gap, timing of spark, and volume of mixture inhaled. And this could only be due to a more effective spark.

The higher voltage of these special coils (and the same remark applies to modern magnetos designed to replace standard coils) calls for some care in the arrangement of the high-tension wiring. It will be appreciated that increased voltage means more liability to leakage of current, and there is little doubt that in production engines the voltage is held down to a figure which, whilst ensuring a satisfactory spark, will not cause undue leakage even when the current-carrying parts are covered with oily filth and moisture. Providing the wiring has been attended to as described in Chapter 3, there should be no cause for complaint.

The use of wide plug gaps will, of course, encourage any leakage and the gap recommended by the coil maker should not be exceeded. It may be found in fact that a gap somewhere between that used with the normal coil and that specified for the 'special' will give the most satisfactory results.

Low-tension matters

The extra coil output naturally calls for more low-tension current to be dealt with by the contact-breaker points. Some makers of high-voltage coils market toughened points and more robustly insulated condensers for use with their product, and these should certainly be included in any change over. Otherwise, the low-tension side does not call for any particular attention since, of course, the voltage remains as before.

After all this, the reader will want to know whether the expense is worthwhile. The answer depends very largely on the type of ignition system fitted as standard. It will be appreciated that in recent times, quite utilitarian cars have been improved in this respect, and wide plug gaps, in conjunction with the appropriate type of coil, are becoming much more usual. When the plug gap as standard is in the region of 0.020 in., it is quite probable that a power increase will be obtained by fitting a high-voltage coil with plugs gapped to 0.030 in. But if the standard set-up includes plugs gapped at 0.025 in. or over, the result of such a change is much less predictable. Further, power increases obtained by using wide-gap plugs (whether as standard or as the result of a 'high-voltage' modification) are felt generally at high revolutions. Drivers who are not particularly interested in more power at this point may be unimpressed since, if the accent is on top-gear cruising rather than quickness in the intermediate gears, the standard parts will probably be adequate.

Resetting of plugs

Using a special coil, the spark plugs will probably call for a little more attention in periodical resetting of their gaps and may not last quite so long as normally. There is, however, no advantage to be gained in using plugs of a different grading from the standard recommendation. The insulators must be kept clean externally, to avoid leakage between terminal and body. When effecting the change of coil it is sometimes feasible, and in fact necessary, to improve on the standard mounting. The Runbaken Oilcoil, for example, must be mounted upright in its special cradle. In deciding where to position the component, it is always advisable to choose a spot as remote from heat as possible. The coil gets quite hot enough in performing its duties as a transformer of voltage without having external heat added to it. When repositioning, it may be found that an extra long high-tension lead is required from coil to distributor. Care must be used in locating this to ensure that it does not come in contact with hot engine parts.

Having discussed special types of ignition coil, it will be useful to continue by considering other advertised aids to more efficient ignition.

Sparking plugs do not come within this category. We have already mentioned the essential point of adhering to the maker's recommendation in regard to plug type, or if a change of make of plug is desired, of accepting the plug manufacturers' ideas on the subject.

We can consider the fittings available for attaching to the high-tension leads, sometimes in the form of outsize plug connectors and which are claimed to 'intensify' the spark in some manner. There is, of course, almost invariably a basis of truth in the claims made for any 'gadget', and it is unfortunate that as competition becomes keener the claims are apt to be exaggerated. In the case of these aids to spark efficiency, a typical form is a well-designed component using a very old idea—that of the auxiliary spark-gap.

Spark-gap terminals

An old dodge, when an engine oils up one of its plugs, is to remove the lead from that plug, hold it about $\frac{1}{8}$ in. from the plug terminal and rev up the engine on the remaining cylinders. It will many times be found that the offending plug will clear itself and recommence firing. As a matter of fact, the idea worked better in the old days of smaller plug gaps and of lubricants which were more easily converted to carbon, than it does nowadays, but it is still worth trying when in a hurry, always providing the operator does not mind receiving the odd shock if he inadvertently touches bare metal instead of insulation.

The artificial spark-gap created in series with the proper plug gap, affects the high-tension current in a manner which induces it to break down the globule of oil bridging the plug points. What it actually does in this regard is a matter for higher electrical theory and need not concern us. The next practical step, however, having found a cure for oiled-up plugs, was prevention, and the outcome was the permanent spark-gap, made by dividing the high-tension leads and inserting in the circuit a small button (not metal, of course) in which the bared ends of the leads were twisted. The danger of having a fusillade of sparks under the bonnet was, of course, obvious, but as a cure for oily engines the idea was quite well known many years ago. Today, the spark-gap can be obtained embodied in an insulated terminal for attaching to the high-tension lead to each plug. It will probably rectify persistent cases of oiling-up which would be far better dealt with by the obvious remedy of reboring or attention to piston rings. The external gap being enclosed there is, of course, no danger of fire as there was in the case of the aforementioned makeshift.

Apart from its merits as an oiling-up cure, however, there is no reliable evidence to show that the device has any other virtue. It should also be borne in mind that any addition to the gap over which the spark has to jump puts an additional stress on the insulation material of the coil.

Spark-gaps are just one item which are to be found offered in this appealing guise. Other devices for fitting in a similar manner have included small choke coils and microscopic condensers. Unlike the spark-gap, which does at least help in one respect, the latter do not appear to have any particular virtue.

Mention of 'other devices' reminds one of the much maligned suppressor, and the story about the individual who, after fitting a suppressor, could not start his engine, the happy ending being, as the reader has probably guessed, that on removing the offending device, everything was fine. While not being in the least concerned with the question as to whether the motorist ought to pay for the convenience of the television enthusiast, it is important not to add to any feeling of being got at financially, the suggestion that engine performance is marred. Unless, of course, there is no circuit through the suppressor (as was probably the case with our friend cited above), performance must be unaffected.

Suppression is nowadays built-in to the ignition wiring in any case, by the use of a high-resistant conductor; if however a supression device has to be fitted say to an older car, there need be no hesitation on the score of reduced power.

Distributor details

Recent innovations have been distributor covers, designed to replace the standard ones, but made of alternative material, in one case transparent. There is no doubt that any improvement in the degree of insulation in the ignition system is bound to be beneficial. Materials, which, apart from their inherent insulating properties, are of a formation which retains these properties under conditions of heat, moisture, and dirt have additional value. It will be obvious that with the tremendous number of plastic-base insulating materials now available, there is considerable scope for experiment, and the electrical manufacturers are fully alive to such developments.

Special types of condenser are of value, as already mentioned, when high voltage coils are used. Otherwise the standard article performs quite satisfactorily. (It is probable that more condenser failures are caused by careless handling than from internal faults.) The same remark applies to contact-breaker points, though less frequent trimming is necessary with some special points, than with the standard article; the gap setting remains correct over a longer period.

So much for the possibility of helping along the ignition by the addition of purchased extras.

We can now do likewise in regard to the carburation side of the power unit.

The carburetter has from time immemorial come in for more than its share of ideas aimed at improving its performance. Its very character makes it liable to this attention, the slight air of mystery about its functioning, coupled with the undeniable fact that few, if any, carburetters produce a perfect mixture all the time, giving encouragement to 'improvers' of one sort and another.

Extra air

As mentioned in Chapter 10, probably one of the earliest marketed additions to the carburetter was the extra-air inlet. Its regular appearances on the market, sometimes in quite complicated form, usually coincide with a period of fuel shortage, when the emphasis is naturally on its possible merits as a petrol-saver. However, in normal conditions of driving, and with everything in the right adjustment, it is doubtful if the addition of a further automatic device will be of benefit, except in very special circumstances.

Mixture agitators

What we like to think of as a fuel-air gas proceeding along the induction pipe is actually a column of air containing fuel drops which attempt more or less successfully to accompany it. Hence, some attention has been given to fitting in the manifold a form of agitator, screen, or other device designed to induce the liquid and air to become more harmoniously united.

Two of these come to mind. One comprises a form of metal washer for sandwiching in the carburetter flange, and carrying a central spindle on which is a small propeller, the tips of the latter just clearing the interior diameter of the induction pipe. The idea is that as the mixture flows past, it rotates the propeller. The second gadget is not dissimilar, but indicated that the inventor decided that having the mixture rotating the propeller might be doing the thing backwards. He therefore arranged for the propeller to remain stationary, and for the mixture to revolve, the washer being provided with fixed and shaped vanes to achieve this end.

There is substance in ideas working on such principles, but the fact is that any improvement will only be marginal. The addition of any fitting which does not show a clear advantage, while appealing to some enthusiasts, introduces extra complication to little or no purpose, and this is undesirable as far as the average driver is concerned.

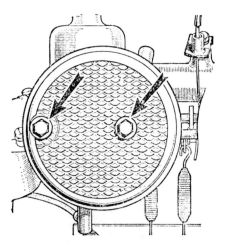

Fig. 11.1. Simple oil-wetted gauze air filter, showing its fixing bolts.

Intake silencers

The very great majority of engines are provided with some form of air intake filter which also serves to diminish the noise of the entering air. This noise was at one time tolerable, not only because the rest of the mechanism provided a background, but also for the reason that it was more constant in tone. Nowadays, with the very wide speed range of engines, and the continual throttle movement in many circumstances, the varying degree of sound emitted can be a nuisance to the occupants of an otherwise quiet motor-car.

Modern types of air filter-cum-silencer perform their job efficiently. The resistance to passage of air is negligible, and in the average engine, the presence or otherwise of the filter has no effect on power output. Some people, however, are ready to aver that they obtain a marked increase in power if the filter is removed. Certainly, the greater amount of noise may create this impression, but if the filter is of a recognised make, and cleaned at the stipulated intervals, its removal cannot possibly have any such effect. It must be emphasised that in the case of highly tuned engines where all steps have been taken to obtain the highest possible power output for some specified purpose, by modifications to such items as induction systems and so on, an absolutely free intake, through a suitable design of 'trumpet', is an advantage. This is because it can be made to tie up with the rest of the modifications to give an overall power increase, on the basis of every little helping. This, however, in no way affects our principle.

Special intakes

In considering the merits or otherwise of various types of air filter it is perhaps a little unfortunate that car manufacturers in some cases have a habit of adopting a different filter design at rather frequent intervals, and having got us used to the idea that, say, the oil-bath type is the only possible one for a particular engine, we are faced with the fact that a little later it is dropped for a gauze affair of rather primitive aspect. For ideas on filter design, readers cannot do better than look at types where efficient filtration is literally a life-and-death matter as, for example, on aircraft and armoured fighting vehicle engines. Admittedly these filters are probably of costly construction, but they do perform efficiently and with absolute reliability. It is probable that some filters recently seen on car engines will require somewhat frequent cleaning if they are to remain unobstructed, whereas such types as, for example, the oil-bath centrifugal variety, though somewhat bulky, do not call for attention very often, at any rate in this country. The oil-bath type of filter is shown on Fig. 11.2. Oil-wetted gauze types, and also filters having a 'throw-away' paper element, are also very effective and do not take up much room (Figs. 11.1 and 11.3). Often the filter body is provided with an intake pipe arranged to draw air from a relatively cool part of the under-bonnet area. This leads to the idea that it might be better still if the air is ducted from outside, at atmospheric temperature. Cold air of course contains a greater weight of oxygen in a given volume than when it is heated, but in this connection, our remarks regarding highly-tuned engines apply once more. The normal touring engine, with its induction system designed for quick warming-up and provision of a homogeneous mixture,

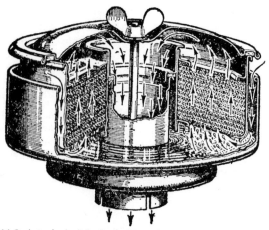

Fig. 11.2. A typical oil-bath air-filter. Arrows show flow direction.

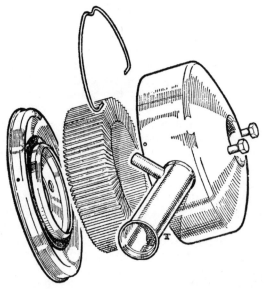

Fig. 11.3. Renewable-element dry-type air-filter with extended inlet pipe to casing.

applies quite a lot of heat to this region deliberately, and the temperature of air entering the filter intake will make little or no difference unless a lot of other drastic modifications (none of which will enhance reputation as a 'touring' engine) are made at the same time. On the contrary, there is merit in having the ability in winter to swing the air intake 'snout' towards the exhaust manifold so as to draw in the hottest air available, in order to counter any tendency to carburetter icing. The 'summer' position reverts to a source of cool, briskly moving air.

Alternative filters

Replacement of the existing filter by an alternative design is a matter for individual choice. As already stated, even the manufacturers in some cases seem unable to make up their minds which is best for their engine. Most designs are based on sound reasoning, which is set out in the specification details pertaining to any particular type of filter. The main thing is to avoid being taken in by fallacies. The use of large external air-intakes on some racing cars, and repeated references to 'ramming' of air into the engine as some magic way of increasing power output, has to some extent influenced filter design. It must be emphasised that having a filter 'mouth' pointing in the direction of travel of the car will not have any effect on performance, impressive though it may look. The long pipes found on some quite 'touring' air cleaners, as on Fig. 11.3, are fitted for reasons of acoustics rather than to gain any ramming effect.

Carburetter changes

British makers generally fit one of four makes of carburetter. It says much for the enthusiasm that motoring engenders even in these hum-drum days (amongst the drive-to-work fraternity as opposed to the sporting side), that the choice of carburetter provides food for many an argument, usually based on mileage-per-gallon figures.

The engine-maker has good reason for deciding on any particular carburetter, his choice being governed mainly by knowledge of the use to which the average owner will put his vehicle. Providing the carburetter tuning, jet sizes and so on, are not interfered with except by an expert (and then only for a very good and unusual reason), that carburetter will enable the engine to give of its best. Any idea that something will be gained by changing the type of instrument is just another fallacy. Here again, the question of a change of carburetter combined with wholesale alterations to the power unit is not covered, as this is a different story.

Exhaust pipe fittings

A wide variety of fittings is available for attaching to the exhaust tail-pipe end. These are often moderately priced, well-finished, and give the car an individual and sometimes improved appearance by

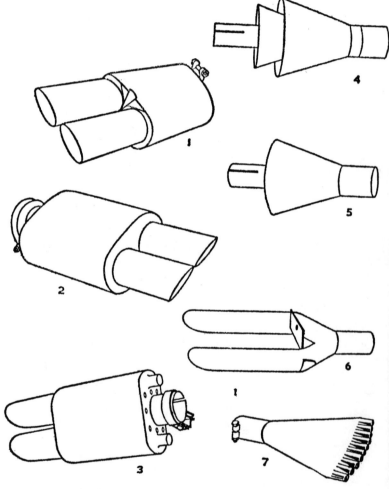

Fig. 11.4. Typical collection of tail-pipe end fittings.
1 Peco patented booster. (S) 2 Aquaplane twin-exhauster. (C) 3 Cornette booster. 4 Turbo-vac extractor. 5 Cornette ejector. 6. Carcraft double booster. 7 Tubex fishtail.

disguising the few inches of rusty pipe which is too often seen even on the most modern vehicles. From the appearance aspect it is obviously not difficult for the owner to make up his mind. When, however, claims are made for improved performance as well, a good deal more investigation is indicated.

In these cases the improvement is usually stated to be the result of pressure-wave damping, auto-extraction of the exhaust by movement of the air-stream under the car, gas temperature reduction, mixing of exhaust and air, or a combination of several of these. It should be made clear that really improved scavenging of the cylinders can result only from bettering the exhaust manifold design; it is at this end of the exhaust system that any scientific approach to pressure-wave action and auto-extraction must be applied, and the further down the pipe we proceed, the less significant will any alteration become. Complete modification of the exhaust system is outside the scope of this book, but readers who are interested in this aspect of design will find further information in a companion book *Car Performance and the Choice of Conversion Equipment.*

A variety of tail-pipe end-fittings is shown on Fig. 11.4, and it needs little discernment to decide which makers genuinely believe that they are offering an article capable of doing a useful job. In view of the moderate cost and ease of fitting of the majority of tail-pipe attachments, the best way for the individual to make up his mind is to try one.

CHAPTER 12

Emission Control Systems

Sealed breathing—Exhaust emmissions—Clean-Air system—Manifold-Air-Oxidation—Duplex manifold system—Other systems—Evaporative emissions

In the pursuit of efficiency, which is the source of both power and economy in a petrol engine, it might be thought that developments aimed at eliminating the emission of unburnt fuel and other pollutants would be beneficial. Unfortunately, however, emission-control systems in general have the opposite tendency—engines fitted with them lose out on both power and economy compared with those innocent of such encumbrances. On the other hand, the mechanisms involved are precision jobs which will respond to meticulous tuning of the calibre which has been advocated elsewhere in this book. Such attention will at least ensure that the minimum of power is being thrown away, while salving one's conscience in complying with the law and making a personal contribution to the reduction of atmospheric pollution.

The motor car was first identified as a major source of pollution (or at least as one which might respond to corrective action) in California in the 1950s. Los Angeles suffers from a peculiar geographical situation which encourages smog to remain stationary in and above the city. The petrol engine was found to exacerbate this natural phenomenon with unburnt hydrocarbon fuel and carbon monoxide, emanating from the exhaust pipe and crankcase breather. Legal limits were imposed, first in California, then the whole of the United States, and in due course they will doubtless spread to other countries with a dense vehicle population which have a pollution problem.

Sealed breathing

Crankcase emissions were the first to be dealt with. It is relatively simple to arrange for the crankcase (and also of course for other internal chambers of the engine, *e g* the rocker cover) to be sealed except for a breather, which is led to the carburetter air intake. Thus the engine inhales its own fumes and, incidentally, by avoiding a pressure build-up in the crankcase, tends to enjoy a reduction in oil leaks. Sealed breathing, otherwise known as positive crankcase ventilation (PCV), is now commonplace on cars sold in Europe as well as North America. Routine attention is confined to cleaning or replacing up to three items which may be fitted: first, the flame trap located in the pipe run to the air intake; secondly, the oil filler cap which often has a built-in filter; thirdly, the PCV valve at the manifold end of the breather pipe. It is important to ensure that the pipes are not obstructed by blockage or kinking.

Exhaust emissions

The various means devised to limit the quantity of unburnt hydrocarbons and carbon monoxide in the exhaust gases are rather more complex. Indeed, as soon as the engineers master one set of conditions, they are spurred on to greater efforts by a tightening of the limits, or by the proscription of new 'villains', such as oxides of nitrogen.

Three principal systems are used to deal with exhaust emissions: the 'Clean-Air' system originated by Chrysler in the USA and adopted by Rootes, Triumph, Rover and Ford (for some models) in Britain; secondly 'Manifold Air Oxidation' (Man-Air-Ox) employed by General Motors and Ford (USA) as well as BMC and Ford of Britain; thirdly the Duplex Manifold system developed by the Ethyl Corporation in America and Zenith in Britain, Jaguar being the most notable British user.

Clean-Air system

The Clean-Air system achieves a 'clean' exhaust by means of precise (and somewhat unorthodox) settings for the ignition and carburetter, with the minimum of additional components. Some engines have had to be detuned in respect of compression ratio and valve timing, but the principle is one of prevention rather than cure —in other words efficiency of combustion is the primary aim.

The Stromberg type CDSE (Fig. 12.1) is typical of the carburetters used with this system. It differs from the normal type CD described in Chapter 4 in having a non-adjustable jet and a needle with a spring-loaded bias; these control the mixture strength during normal driving to the precise limits required, assisted by a temperature compensator, which senses the variations of underbonnet temperature and adjusts an extra air bleed accordingly. Overrun conditions tend to produce high emissions, since the mixture in the cylinders is diluted with exhaust gases and combustion is far from complete. To overcome this, a by-pass valve is opened by manifold

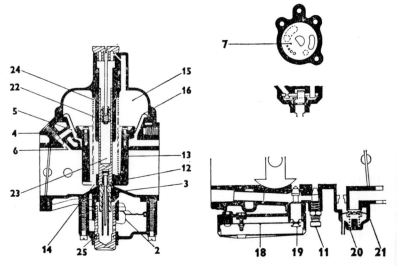

Fig. 12.1. Zenith-Stromberg special emission control carburetter, type CDSE.

2 Twin floats
3 Jet orifice
4 Leak balancing screw
5 Duct
6 Duct
7 Starting disc
11 Idle trimming screw
12 Biased jet needle
13 Air valve
14 Duct
15 Diaphragm
16 Depression chamber
18 Bi-metallic blade
19 Tapered plug
20 Throttle by-pass valve
21 Diaphragm (by-pass valve)
22 Damper
23 Hollow guide rod
24 Coil spring (air valve)
25 Rubber 'O' ring

vacuum and passes extra combustible mixture to the engine; extra power output would result, which is not wanted during deceleration, but this is counteracted by a vacuum-retard capsule on the distributor which retards the ignition by several degrees.

The only adjustments available to the owner are normal idling speed and fast idle speed; the former is controlled by the throttle stop screw, the latter by its own screw. Fortunately, the precise limits to which the carburetter is manufactured remove the necessity for 'tinkering', as faulty running will generally be found to be due to ageing sparking plugs, contact points, valves or some such commonplace malfunction. Twin-carburetter synchronization is as important as ever, of course, and so are such things as oil in the dashpots, lack of air leaks at the manifold flanges and a reliable supply of fuel from the pump. The special emission components seldom give trouble that can be detected during normal driving.

Manifold-Air-Oxidation

Like the Clean-Air system, 'Man-Air-Ox' uses a precision-tuned carburetter, of conventional design except for a spring-loaded valve in the throttle butterfly to pass combustible mixture on the overrun. Ignition settings are usually not 'messed about', but there is a special accessory fitted to the engine in the shape of an air pump, belt-driven from the crankshaft. Air is pumped via a non-return valve and a manifold to the exhaust ports, so that its oxygen content will combust the unburnt hydrocarbons and carbon monoxide in the exhaust into innocuous carbon dioxide. Engine power is required to drive the pump.

Fig. 12.2 shows a typical installation, including a device with the intriguing name of 'gulp valve'. Its function is to admit a 'gulp' of

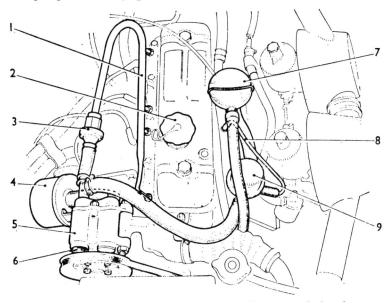

Fig. 12.2. Manifold-Air-Oxidation system installed on a typical engine.

1 Air manifold
2 Oil filler cap with filter
3 Non-return valve
4 Air cleaner for air pump
5 Air pump
6 Relief valve
7 Crankcase breather control valve
8 Vacuum connection to gulp valve
9 Gulp valve

air from the pump into the inlet manifold when the throttle is closed, to prevent the mixture becoming momentarily too rich.

With all these extra 'tappings' into the inlet and exhaust systems, it is clear that there must be no leaks through any of the hoses or connections. Either leaks or a faulty gulp valve can cause problems such as backfiring, hesitant acceleration after a sudden closing of the throttle or apparently unstable carburation. Tension of the air pump drive-belt should be checked periodically.

Duplex manifold system

Features shared with the other systems already described include precision-tuned carburetters with throttle by-pass valves and a distributor with a vacuum retard capsule. The special feature is the dual throttle and dual inlet manifold arrangement, shown on Fig. 12.3. During idle and part-throttle operation the main throttles remain tightly shut, and all mixture passes through the auxiliary ones and their swirl chamber, mounted on the exhaust manifold. The gas flow and exhaust heat ensure that the fuel is thoroughly atomized before it reaches the engine, to maintain the optimum air:fuel ratio. At wide throttle openings, the main butterflies are operated as well, and gas flow for maximum power is unimpeded.

The principal user of this system, Jaguar, has found it necessary to add a refinement consisting of an automatic hot- and cold-air mixing valve which maintains the air fed to the carburetters at a steady temperature.

Other systems

Fuel injection offers apparent advantages over carburetters for emission control, in the way of precise fuel metering and equal distribution among cylinders. Modern design can provide sensors to vary the mixture in response to virtually any parameter that can be envisaged and so provide for complete combustion in all circumstances. However, few manufacturers have adopted fuel injection specifically to combat emission—and one or two have actually reverted from injection to carburetters for their North American models. A notable exception is VW, who fit an electronically-controlled fuel injection system, claimed to give a power bonus as well as a clean exhaust.

Another approach to the problem is a chemical one. So-called 'catalytic' silencers are used, containing exotic materials which act as catalysts with the hydrocarbon and carbon monoxide content of the exhaust gas. Although very low emissions are claimed with

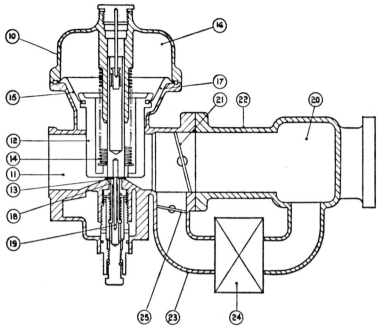

Fig. 12.3. Zenith duplex manifold system.

10 Dashpot cover
11 Air intake
12 Air piston
13 Jet housing
14 Piston spring
15/17 Diaphragm
16 Dashpot
18 Jet
19 Jet needle
20 Menifold
21 Main throttle
22 Main throughway
23 Auxiliary duct
24 Heated swirl chamber
25 Auxiliary throttle

these devices, they are still very expensive and of unproven durability.

Evaporative emissions

Even the petrol in the tank and in the carburetter float chambers gives off toxic fumes—in small quantities—as it slowly evaporates; regulations are already in effect in America to subdue even these. It is easy enough to vent the petrol along to the air cleaner, in parallel with the sealed crankcase breather, although a suitable flame trap is desirable. This, of course, works only when the engine is running, whereas fuel gives off its vapour all the time. The likely solution seems to be a canister in a line to the tank breather containing a substance such as activated charcoal, to absorb and store the petrol fumes until they can be sucked away when the engine is run again.

Chapter 13

Power to the Best Advantage

Legitimate power absorption—Saving power—Brake adjustment—Useful tests—Test-tune equipment—The equipment—Interpretation of readings—Gauge readings—Conclusion

Having tuned an engine to give of its best, it has to be kept in mind that power at the flywheel if it is to be of practical use, must be transmitted to the road-wheels; in its passage thereto, it is surprising how much of it can be lost. Some loss is, of course, inevitable, but there is no excuse for this to be excessive.

Legitimate power absorption

As is generally known, the bulk of the power available is used in pushing the car through the air, and modern coachwork designs have tackled this aspect very thoroughly, to the benefit of performance and economy. This problem is, in any case, a matter outside the control of the owner. He can, however, in many cases do something about the other items which absorb power, such as rolling-resistance and transmission friction.

Tyre pressures, and choice of tyre tread, can have considerable bearing on the first-named. The recommended pressures should always be adhered to, and it is frequently worth while trying a pound or two in excess of that figure, which procedure may be found to improve handling as well as economy. The question of tread pattern hardly arises if the tyres are of a reputable modern design. It cannot be emphasised too stongly that odd 'bargain' tyres must never be used. Road conditions these days demand absolutely reliable tyres which have good wearing qualities and which grip the road firmly in conditions of heavy breaking in emergency on wet roads. Tyres are certainly not cheap, but this is one case of money well spent.

Regarding transmission friction, the difference between the power developed at the engine flywheel and that actually available at the road-wheels can be unbelievably high. This is obviously a loss that has to be accepted, but even with everything as it should be, the transmission may account for 20 per cent of the engine power. Such defects as rubbing brakes will make this considerably more. Taking a 100 b h p engine, up to 20 b h p seems a lot to give away, and it will be queried where this power goes to. First there is the churning of oil in the gearbox and axle. Next, the frictional loss between meshing gearwheels which will vary according to the gear in use. The universal joint angularity adds something to the total, and finally there is friction in the various bearings and oil-seals, oil-drag in the axle casing, and windage loss caused by the 'fanning' effect of the revolving parts, including the road wheels.

Saving power

The references to oil-drag above-mentioned are important. The answer to mitigation of this loss is to use always the recommended grade of lubricant. Do not be tempted to use 'thicker' oil because it gives a quieter axle or gearbox, or because leaks are troublesome. Apart from increasing the power loss, thicker lubricant may actually starve some important component. Noisy gears are very often not doing any harm, but if troublesome can be rectified by overhaul. Oil-leaks are usually the result of defective seals, which can be replaced. These are, of course, usually jobs for the service station.

Too much grease in wheel bearings also increases drag and becomes troublesome in hot weather, when it is liable to exude on to brake linings. When grease nipples are provided on hubs, do not be tempted to pump in more than about three strokes at 6,000 miles intervals; overfilling does far more harm than good, as this form of lubrication lasts almost indefinitely.

Bent axleshafts, propeller shaft and wheels can all absorb power, but it is naturally assumed that no self-respecting owner will run a car with these defects. We can therefore go on to what is a very common source of unnecessary power loss—that of rubbing brakes.

Brake adjustment

With the older types of mechanically-operated braking systems, adjustment usually takes time and requires some care; it is also necessary to keep the mechanism properly lubricated and free from wear. With modern hydraulic systems, the simplicity of layout and absence of control cables or rods has led to the adoption of extremely simple means of adjustment, enabling the brake-shoe clearance from the drum (in theory at any rate) to be set correctly in a matter of minutes.

In actual fact, to do the job properly takes quite a time, and requires the help of someone to operate the brake pedal, if the right balance between minimum lost motion and no rubbing of shoes is to be obtained.

Earlier types of typical hydraulic brakes have individual adjusters for each shoe, the 'business' end of the adjuster comprising a hexagon head protruding from the back-plate. Inside, this carries a cam which bears on a stop-pin fixed to the shoe. Rotation of the hexagon in its self-holding bearing bush brings the cam against the stop, the shape of the cam moving the shoe against the drum when it is turned. The arrangement is shown on Fig. 13.1.

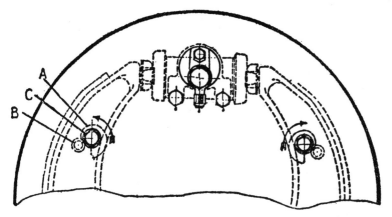

Fig. 13.1. Early type of brake adjustment using a cam and stop on each shoe.
A Cam inside backplate. B Stop on brake shoe. C Adjusting hexagon head on outside of backplate.

For adjusting, the usual instruction is to jack up the wheel, spin it, and turn the hexagon until the shoe is just heard to rub on the drum. The cam is then backed off until the shoe just clears, and all should be well. Actually, the snag is that the rubbing may be caused by grit in the drum or a minor high-spot on the brake lining. For this reason, an assistant operating the brake pedal is invaluable. The procedure then is, that after spinning the wheel, and before attempting any adjustment, the pedal is applied hard, to an extent which locks the wheel. This is done several times, to iron out any high spots and mitigate the effect of grit. Then carry out the adjustment, applying the brake and testing the clearance in between small movements of the hexagon adjuster. It will be found by this means that eventually the shoe will clear with a very small degree of lost motion. The procedure should, of course, be carried out on all shoes, and all separately.

Later types may have a hole in the brake-drum which gives access to a screwdriver-slotted 'clicker' adjustment as shown on Fig. 13.2.

Fig. 13.2. Brake adjustment by access holes in drum for screwdriver-operated 'clicker' adjusters.

The usual instruction is to turn the adjuster clockwise until it can be turned no more, when the shoe will be hard on, and then back it off one or two 'clicks'. The snag in this case is similar to that already mentioned, *i e*, that one or two clicks may not allow the shoe to clear entirely. Here again the same procedure of pedal actuation is followed, when it will be found that a satisfactory compromise can be arrived at.

Another method of adjustment similar to the foregoing has the adjusters in the form of square shanks protruding from the brake back-plate as shown on Fig. 13.3. Each adjuster is turned as required by a spanner, but otherwise the procedure is the same as already described.

Self-adjustment is one of the advantages of most disc brakes, a typical design of which is shown on Fig. 13.4. As the friction pads wear, the hydraulic pistons in the caliper follow them inwards towards the disc, retracting only a short distance after each brake application. The routine attention which must not be neglected is inspection for pad wear and topping-up of the fluid reservoir.

It is important to note that on hydraulic systems there must be no bar to the free return of the master cylinder piston when the brakes

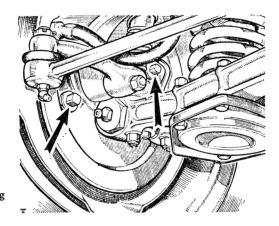

Fig. 13.3. Squared heads on outside of backplate for operating 'clicker' adjusters.

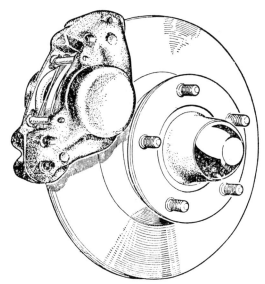

Fig. 13.4. Disc brake assembly with quickly-detachable friction pads.

are released. The pedal must at all times be kept free on its pivot and a small amount of lost motion must always be present at the pedal pad—about half an inch is usually sufficient—to ensure that the master piston comes right back in its cylinder to give communication between the latter and the fluid reservoir. (See Fig. 13.5.) Some brake pedal pivots lack any means of lubrication and in such cases, light oil may help to keep everything free. Earlier types of self-lubricating bushes also were somewhat unpredictable in locations where road dirt and water are present, and additional oiling does no harm. This is not necessary on the present-day bushes of non-metallic or self-oiling type.

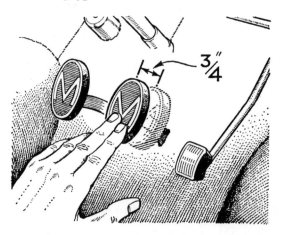

Fig. 13.5. Free brake pedal movement, as shown, may be specified to ensure that the shoes clear the drums.

Useful tests

A useful check on brake freedom is to feel the drums occasionally (not, of course, after heavy braking) to ensure that they are reasonably cool. Some slight heat transmitted from the axle is, of course, to be expected at the rear, but this is easily distinguisable from that produced by rubbing shoes.

When it is found that the adjustment becomes really difficult in obtaining a nice balance between minimum lost motion and shoe rubbing, it is probable that the shoes need relining.

Other hydraulic systems, and also those mechanically operated, can be adjusted on similar lines. In the case of the latter, it is important to ensure that there are no stiff joints or cables in any part of the system, and that all pull-off springs are in good order; here again, the presence of the 'pedal-operator' is a great help.

Finally, if one judges how much effort is required to push the car on a level garage floor, with tyres at the correct pressure and everything else as it should be, it is quite easy to determine whenever any untoward friction is developing, so that action can quickly be taken, even though this may mean nothing more than blowing up the tyres to the right pressure, or freeing-off and lubricating the handbrake linkage, which commonly sticks in the 'on' position.

Test-tune equipment

A great many garages are nowadays equipped with intriguing-looking and expensive apparatus whereby the correct adjustments to an engine can be made in a very short time without taking the car on the road, and (in theory at all events) with a precision denied to human frailty. However, on occasion an owner has found that after obtaining by painstaking attention of the kind we have described, a highly satisfactory performance, a final check by the garage tester, using the apparatus, shows various faulty adjustments, pinning these down with remarkable exactitude. After readjusting as prescribed to obtain the results specified for maximum performance, he is then disappointed to find that the latter is rather worse than before the check was done. The question naturally arises—are his rule-of-thumb tuning methods, while apparently effective, giving perhaps an illusion of extra power which is actually harming the engine; in other words, is it best to follow strictly the findings of the test-tune plant and be content with an inferior performance which is apparently proven to be correct?

The answer is to use intelligent discretion. Many of the readings obtained on these machines are extremely valuable, but on the other hand they depend on intelligent interpretation so that the human element can still eventually intrude. Ultimately, however, road performance is the thing that matters, and many times this is apt to confound 'what should happen' in theory.

The equipment

Modern garage testing equipment is extremely valuable as an aid to diagnosis of an engine's ills. Earlier types seemed to bear some relationship to fun-fair machines, in their desire to attract the motorist, but this phase is now past, and the equipment can be seen for what it is. That is, a collection of standard meters which when suitably connected to the engine, register various readings. Considering any basic engine, we know that fundamentally its operation depends on variations in pressure. We also have an apparatus for supplying fuel to the carburetter, and an electrical system providing power for ignition and auxiliaries. So long as all these functions are being correctly performed, the engine will be working at a high state of efficiency.

For obtaining complete evidence of the well-being or otherwise of the engine, we therefore require appropriate meters which will measure the following: induction manifold vacuum; cylinder compression pressure; fuel-pump delivery pressure; lubricating oil pressure in the main gallery; oil pressure in auxiliary supply, $e\ g$, to rocker gear, etc.; water temperature; exhaust port temperature; generator output voltage and current; generator field voltage and current; battery voltage; ignition timings; ignition insulation resistance; condenser insulation resistance; oil temperature.

Interpretation of readings

If we imagine that a set of gauge readings approximating to the foregoing have been obtained, where do we go now? Consider one or two. The carburetter may be flooding and the first advice might be, renew the needle valve and seating. But the test gauge might well show too high a fuel-pump delivery pressure, and rectification of this would obviously cut out wasted expenditure on carburetter parts. It might be objected that the pump pressure could hardly increase during service, but an incorrect replacement could cause this.

The vacuum gauge coupled to the induction manifold is quite possibly the most informative of the lot, and for this reason can be obtained on its own for instrument panel mounting, providing a permanent tell-tale of engine condition. Though in this case the dial is sometimes embellished with rather colourful descriptions, the plain vacuum gauge which does the same job is calibrated from zero to 15 lb per sq in. vacuum, or 30 inches of mercury ditto. The pointer will read zero at atmospheric pressure and it reads 'backwards' in comparison with a pressure gauge, as it measures pressures below the normal air pressure, not above.

Gauge readings

With the throttle almost shut and the engine idling there is a high vacuum in the induction manifold so that a steady reading on the gauge of 18 to 20 inches of mercury (or 9–10 lb negative pressure) should be obtained on a good engine. When the engine is revved up and the throttle snapped shut, there will obviously be a high vacuum at the instant of throttle closing and a momentary reading of 25–30 inches will be seen. Lower readings when idling suggest defective pumping action so that piston rings, bores or valves must be suspect. However, an engine with appreciable valve overlap may give slightly erratic readings, due to varying exhaust pressure, but these should be consistent and not widely fluctuating.

Oscillation of the gauge indicates weak valve springs or sticking valves, and in such cases the oscillations increase with engine speed. Late valve timing (which would, of course, be a major maladjustment when overhauling) or unduly rich mixture, will also cause a slow oscillation, the latter usually being obvious also from other symptoms.

A pressure gauge applied to each cylinder bore in turn, via the spark-plug hole, while turning the engine with the starter, will confirm bore condition and also valve tightness and gasket condition. Consistency of readings as between individual cylinders is an excellent sign of good condition, while any marked discrepancy points to impending trouble such as a valve burn-out.

As far as the electrical system is concerned, the tests are those that would normally be carried out by any skilled craftsman; indeed it is impossible to track electrical faults in a methodical way without them. The condition of the water jacket can be shown by taking readings of exhaust port temperatures and again such readings should be reasonably consistent, otherwise the indication is of blocked water passages in the area having the 'high' reading.

It will be seen that, shorn of its frills, and intelligently used, the test-tune equipment can provide a lot of information in competent hands. For those who take an interest in their power units, and take advantage of these facilities, it is advantageous to log the information for future reference, and also to make friends with the operator and put a few discreet questions to him as to the conclusions reached, and why.

Conclusion

Apart from the brief outline of special accessories given in Chapters 10 and 11, this book has been concerned, not with the super-tuning of an already efficient engine, but rather with the cancelling-out of factors which prevent full efficiency from being obtained. In case any reader may think that what has been done is merely to make rather a lot out of what are quite simple adjustments, he can rest assured that such is definitely not the case. The author has proved this over and over again, on all types of engine, in a period of 45 years.

There is a lot of satisfaction to be had from driving an engine which is known to be 'just right'; not only from the point of view of the performance available, but (less charitable yet quite human) from the envious comments of friends with identical cars that 'won't accelerate'—or start, or idle, or something—'like yours'.

Many modern cars are so much alike that it might be thought impossible to endow one's own vehicle with a personality; owner-attention of the kind detailed, however, will go a long way towards achieving this.

APPENDIX

Explanatory Definitions

On occasion, even the well-informed find it difficult to convey in simple language the meaning of commonplace technical terms. The following definitions may prove useful in such situations.

- HORSE-POWER (h p): The unit of work; 1 h p is equal to 33,000 ft/lb per min. This is the work done in lifting a weight of 33,000 lb through a distance of one foot in one minute, or any other quantities of pounds and feet which when multiplied give 33,000.

- BRAKE HORSE-POWER (b h p): The actual power developed at the engine-shaft, as measured on test by coupling the engine to a dynamometer or 'brake' for absorbing the power.

- TORQUE: The effort applied to a shaft or wheel which tends to turn it. When the torque is of sufficient value to rotate the shaft through a definite distance in a given time, work is done, which is then stated in hp. The usual unit of measurement is pounds–feet (lb–ft).

- MEAN EFFECTIVE PRESSURE (m e p), or MEAN INDICATED PRESSURE (m i p): The average pressure produced in the cylinders on the explosion stroke, and which results in the shaft-power. It is measured at the cylinder itself, by a scientific apparatus, or 'indicator'.

- BRAKE MEAN EFFECTIVE PRESSURE (b m e p) or BRAKE MEAN PRESSURE (b m p): A figure analogous to the above, but obtained by calculation from the actual b h p. The figure is imaginary, as it allows for the mechanical losses in the engine. It does however enable useful comparisons to be made between designs, in almost all relevant aspects.

- MECHANICAL EFFICIENCY: A factor expressed as a percentage, which shows how much of the explosion pressure is obtained as power at the engine-shaft; it is the percentage difference between the m e p and the b m e p. The loss is due to friction, inertia, etc., in the engine moving parts.

- THERMAL EFFICIENCY: A factor expressed as a percentage which shows how much power is obtained from the heat energy in the fuel used; it is thus an indication of the effectiveness of the engine as an apparatus for converting heat into work.

VOLUMETRIC EFFICIENCY: A factor showing the degree of completeness with which a cylinder is charged, exhausted and recharged as the operating cycle is performed; it is a measure of the adequacy of valves, ports, induction and exhaust systems, etc.

Car performance

The catalogues issued by reputable car makers are intended not only to show the vehicle in as attractive a light as possible, but also to provide the prospective buyer with the information he needs, in order for him to decide whether the mechanical characteristics are commensurate with his kind of motoring. Manufacturers however are by no means unanimous as regards their methods of obtaining performance data, and this has to be borne in mind when making comparison as between different models in a similar class.

Engine output

Engine output, given as maximum brake horse-power at a specified rate of rev/min, is obtained by coupling the engine to a power-absorbing dynamometer on a test-bed. Some makers when testing in this manner, disconnect accessories such as the fan, water pump and generator all of which, in practice, absorb a certain amount of engine power. Further methods of increasing the power obtained at the engine flywheel are to employ a special test-bed exhaust system which permits a much freer passage for the exhaust gases than is the case when the standard pipe and silencer are in use. The test-house temperature may also be kept at the best figure for maximum output, while cases are known whereby the carburetter and ignition timing have been adjusted to obtain maximum performance at each particular increment of engine rev/min when obtaining the b h p/ torque curves. Performance figures which are obtained in the manner typified by the above, are commonly found in American car catalogues, but several British makers also use the method. The figures are then known as Gross, or SAE, the initials denoting the Society of Automotive Engineers whose headquarters are in New York. All continental cars, and many British ones, carry out their engine tests under conditions equivalent to those when installed in the car, with all accessories driven, and with the standard exhaust system. This is called the Nett figure, or DIN (Deutsche Industrie Norm) in the case of continental engines.

Obviously the nett figure is considerably below the gross, the reduction being as much as 15 to 20 per cent. There is also a minor difference between the nett figures for British and continental engines, as the latter equate 750 Watts with 1 h p instead of the British 746 W. Thus 1 h p is equal to 1·014 metric h p.

Torque and brake mean pressure

The figures given for maximum torque can be related to the acceleration and hill-climbing power of the car by taking in the laden weight and gear ratios. However, this requires a fairly expert appraisal, and it generally suffices to note the engine rev/min at which maximum torque and maximum b h p occur. If the former is obtained at a low engine speed, in comparison with the speed for maximum b h p, the engine will have been designed in general to pull strongly from fairly low speeds. This characteristic is useful in cars which are liable to be heavily laden or used for towing caravans and trailers. The maximum horse-power, however, will be limited in comparison with engines on which maximum torque occurs higher up the speed range. The latter condition indicates that the engine is built for brisk revving and free use of the gearbox, in which case a high performance is obtained. In general, engine design tends towards this type nowadays.

Compression ratio

There has been a great improvement of late both in the quality of fuel and the variety available. Designers have taken advantage of these developments, which enable the compression ratio to be increased with a consequent increase in the thermal efficiency of the engine. Higher torque is also obtained, so long as the compression ratio is not increased beyond the maximum advised for any particular fuel grading. It is the rule nowadays for the car maker to specify the fuel to be used, particularly where alternative compression ratios are offered. In such cases, the lower ratio is usually intended for buyers who have to use the vehicle in countries where lower-grade fuel only is obtainable. For environments where motoring is well established, however, compression ratios of between 8 and 9 : 1 are quite normal on family-type saloons, while high-powered sports car engines may run on ratios from 10 to 12 : 1, when using the highest grades available.

Index

A

AC fuel pump operation   102
— pump dismantling   104
— —, tappet positioning   103
Accessory wiring   142
Adjustable rocker fulcrum   4
Adjustment of starting carburetter   63
—, Solex carburetter   72
Advantages of thin oil   118
Air leakage in wiper   39
— — to manifold   37
Alternator principle   140
Anti-freezing mixture   132
Aural check, carburetter synchronisation   83
Automatic air valve   177
— breather valve inspection   39
— timing control, lubrication   28
Auxiliary starting carburetter   62

B

Battery charging   153
— connections   140
— operation   151
Bearing oil film maintenance   118
— wear and oil pressure   115
Belt tension   139
Bi-starter, Solex   71
Boiling damage   131
Brake adjustment   195
— pedal movement   205
By-pass oil filter circuit   113

C

Calorific value of fuel   170
Carburetter air leakage   76
— joint rectification   40
— spindle wear   40
Castor-base racing oil   118
Catalytic silencers   199
Centralising jet, Stromberg   68
— —, SU   49
Centrifugal timing control   23
Chamber cleaning, SU pump   92
Choice of engine lubricant   117
Clean-Air system   195
Cleaning jets, method   76
— plug threads   16
Climatic conditions   151
Clips for wiring   149
Cold-starting conditions   152
Condenser defects   31
— removal   25
Connections to by-pass filter   129
Contact breaker dismantling   24
— — examination   20
— — faults   19
— — spring testing   31
— —, vintage type   23
— points, alignment   29
— position for static timing   34
Control of starting carburetter   63
Cooling system design   131
Corrosion in distributor   21
Cylinder pressure reading   209

D

Dashpot reassembly, SU   47
— spring grading   57
Defects of SU jet   49
Deposits in cylinder jacket   131
Detachable type plugs, cleaning   15
Detergent oil as a cleaner   118
— —, use   118
Diaphragm-jet adjustments   61
— — type SU carburetter   60
— removal, SU pump   92
— replacement, AC pump   106
Distributor contact cleaning   22
— cover cleaning   21
— modifications   186
— pick-up brush attention   20
— position for static timing   33
Drain plug, engine, position   120
— — security   122

INDEX

— — tightening 121
Driving for economy 169
Dry-element air filter 189
Duplex manifold system 198

E

Earthing of accessories 150
Economical tyre pressures 170
Economy carburetter setting 179
— devices, manifold 174
— setting, carburetter 75
Electric cable runs 147
Electrode gap dimensions 16
— shape 13
Engine idling speed 75
Evaporative emissions 199
Exhaust emissions 195
— pipe fittings 191
— smoke and engine wear 116
Extra air control 176
— air disadvantages 176

F

Faulty carburation symptoms, Zenith 80
Feeler measurement, rocker clearance 2
Filter cleaning, PD pump 100
Final adjustment of rockers 4
— ignition adjustment 36
Fitting a by-pass filter 128
— diaphragm, Stromberg 69
— rocker cover 10
— spark plugs 14
Float level adjustment, Stromberg 68
— needle valve wear 41
Flushing cooling system 132
Ford, old type tappets 5
Friction loss from lubricant 118
Frictional power loss 202
Fuel and power 170
— filter, AC pump 104
— —, SU carburetter 53
— —, Zenith carburetter 81
— injection 198
— level adjustment, SU 55
— metering, SU carburetter 42
— pressure 76
— pump air leakage check 107
— — testing, SU 96
— — valve check 93

Fuse capacity 142
Fuses and wiring 141

G

Gadgets on show 182
Gearbox power loss 202
Generator belt tension 156
— brushes, attention 155
— lubrication 155
— wiring 156
Grommet fitting for wiring 148

H

Heat loss in engine 171
High tension cable removal 20
— — connections, method 22
Hose clip fitting 137
Hot spark effects 183
Hydraulic brake adjustment 203

I

Idling adjustment, Solex 75
— —, Stromberg 67
— —, Zenith 80
Ignition contacts rectification 27
— system cleanliness 160
— timing and fuel 19
— — by road test 35
— —, static 32
Insulating material 146
Insulator leakage on plugs 14
Intake filters 189
— silencers 188
Interpretation of test readings 208

J

Jet needle examination, SU 46
— selection, Solex 73

L

Lag-type throttle interconnection 83
Leakage of water 134
Legitimate power loss 194
Low tension connections 23

M

Main earth connection 140
Makeshift wiring 139
Making electrical connections 147
Manifold-Air-Oxidation 197

INDEX

Manifold clamp tightening 39
— design 173
— flange rectification 38
— joint packing 38
Measurement method, ohc gear 6
Mechanical braking system 206
Mileage per gallon 172
Mixture agitators 187
— correction, Solex 74
Modern high-tension wiring 22
— lubricants, and oil flow 111
Multiple carburetters, synchronising 81

N

Needle valve rectification 54
— — wear symptoms 41
— — — test 41
Non-detachable plug cleaning 16

O

Oil circulation principle 108
— deflector failure, symptoms 9
— —, valve cap 8
— distribution gallery pressure 115
— drag loss 202
— feed to ohv gear 117
— filter assembly 125
— — connections 126
— — location 112
— — overhaul 126
— passage layout 110
— pressure and running conditions 111
— — drop causes 114
— — gauge readings 112
— — valve defects 114
— pump delivery volume 110
— relief valve operation 111
— wetted air filter 189
Overflow water loss 133
Overhead camshaft housing 10

P

PD pump, contact cleaning 101
— —, testing 102
— type electric pump 99
Performance deterioration general 84
Pinking on road test 35
Plug examination and engine faults 14
— gap setting 16
— — with special coil 184
— oiling, remedies 185
— points, cleaning 16
— weather protection 161
Positive crankcase ventilation 194
Power loss percentage 201
— to drive accessories 171
Pre-ignition causes 13
Pressure filler cap operation 133
Principle of SU carburetter 41
Pump contact points, cleaning 98
— gland leakage 135
— valve replacement, SU 93
Pushrod examination 8

R

Radiator cement 166
— leakage 166
Removing distributor wiring 20
Renewable element filter, type 113, 122
Replacing pump diaphragm 94
— SU needle 47
Research by plug makers 18
Restricted oil flow to filter 129
Reverse-flow adaptor 132
Reversing light connection 144
Rich mixture symptoms 58
Rocker cover distortion 10
— — joint 9
— gear oil passages 8
Rotating engine for testing 3
Rotor-arm rectification 23

S

Sealed breathing 194
— cooling system 134
Security of manifold nuts 40
Shim adjustment, ohc gear 6
Side-valve engine tappets 5
Single-bolt filter attachment 126
Six-cylinder engine rocker adjustment 3
Slow running adjustment, SU 57
Soldering material 146
Solenoid starter switch 159
Solex carburetter operation 71
— —, examples of jet selection 74
Spanner for plugs 17
Spark-gap terminals 185

INDEX

Spark plug life 16
— — removal 14
— — temperature 13
— — type 13
Sparking devices, claims 18
Special ignition coils 182
Starter engagement requirements 157
— motor maintenance 157
— pinion unsticking 158
— switch operation 158
— —, type 156
Starting device, Solex 71
Stroke adjustment, SU pump 97
Stromberg carburetter design 64
— — operation 66
SU carburetter variations 43
— control linkage 50, 61
— dashpot examination 44
— — rectification 46
—float, types of 56
—jet, position of components 51
— pump, contact breaker adjustment 90
— — dismantling 90
— — operation 85
— — pressure grading 88
— retarder, maintenance 48
— starting mixture control 59
Sump draining accessories 120
— — routine 119

T

Terminal connections 145
Test-tune equipment 207
Thermostat operation 137
— testing 138
Throw-away oil filter design 127
Top dead centre, finding 33
Trickle charger advantages 154
— — type 154

Triple carburetter tuning 83
Twin carburetter interconnection adjustment 81
Tyre pressures, 170

U

Under-bonnet temperature 180
Undersealing compounds 164
Use of liquid jointing 38
— — thick oil 116

V

Vacuum control inspection 28
— gauge interpretation 208
— timing control 23
Valve accessory check 8
— assembly, AC pump 106
— clearance measurement 1
— — adjustment 3
— gear lubrication, check 7
Vapour locks 168
Vernier adjustment of timing 34
Viscosity of engine oil 118

W

Warming-up and oil flow 112
Water hose defects 136
— ingress protection 162
— overflow bottle 134
— pump lubrication 136
— — shaft seal 136
— temperature 180
Winter disabilities 152
— unreliability 151
Wiring accessories 144

Z

Zenith carburetter principle 77
— — starting mixture 78